Yearbook of
Astronomy 2023

YEARBOOK OF
ASTRONOMY
2023

EDITED BY

Brian Jones

WHITE OWL
AN IMPRINT OF PEN & SWORD BOOKS LTD.
YORKSHIRE ~ PHILADELPHIA

First published in Great Britain in 2022 by
WHITE OWL
An imprint of
Pen & Sword Books Ltd
Yorkshire – Philadelphia

ISBN 978 1 39901 844 9

A CIP catalogue record for this book is available from the British Library.

Typeset in Dante By Mac Style
Printed and bound by Short Run Press Limited, Exeter

Pen & Sword Books Ltd incorporates the Imprints of Pen & Sword Books Archaeology, Atlas, Aviation, Battleground, Discovery, Family History, History, Maritime, Military, Naval, Politics, Railways, Select, Transport, True Crime, Fiction, Frontline Books, Leo Cooper, Praetorian Press, Seaforth Publishing, Wharncliffe and White Owl.

For a complete list of Pen & Sword titles please contact

PEN & SWORD BOOKS LIMITED
47 Church Street, Barnsley, South Yorkshire, S70 2AS, England
E-mail: enquiries@pen-and-sword.co.uk
Website: www.pen-and-sword.co.uk

or

PEN AND SWORD BOOKS
1950 Lawrence Rd, Havertown, PA 19083, USA
E-mail: Uspen-and-sword@casematepublishers.com

Contents

Miscellaneous

Editor's Foreword

The *Yearbook of Astronomy 2023* is the latest edition of what has long been an indispensable publication, the annual appearance of which has been eagerly anticipated by astronomers, both amateur and professional, for well over half a century. As ever, the Yearbook is aimed at both the armchair astronomer and the active backyard observer. Within its pages you will find a rich blend of information, star charts and guides to the night sky coupled with an interesting mixture of articles which collectively embrace a wide range of topics, ranging from the history of astronomy to the latest results of astronomical research; space exploration to observational astronomy; and our own celestial neighbourhood out to the farthest reaches of space.

The *Monthly Star Charts* have been compiled by David Harper and show the night sky as seen throughout the year. Two sets of twelve charts have been provided, one set for observers in the Northern Hemisphere and one for those in the Southern Hemisphere. Between them, each pair of charts depicts the entire sky as two semi-circular half-sky views, one looking north and the other looking south.

Lists of *Phases of the Moon in 2023* and *Eclipses in 2023* are also provided, together with general summaries of the observing conditions for each of the planets in *The Planets in 2023* and a calendar of significant Solar System events occurring throughout the year in *Some Events in 2023*. Apparition charts for all the major planetary members of our Solar System have been compiled by David Harper. Further details are given in the article *Using the Yearbook of Astronomy as an Observing Guide*, the apparition charts themselves following the article *The Planets in 2023*.

As with *The Planets in 2023* and *Some Events in 2023*, the *Monthly Sky Notes* have been compiled by Lynne Marie Stockman and give details of the positions and visibility of the planets for each month throughout 2023. Each section of the *Monthly Sky Notes* is accompanied by a short article, the range of which includes items on a variety of astronomy- and space-related topics including *The Incomparable Sir Patrick Moore*, written by Neil Haggath to commemorate the centenary of the birth – in Pinner, Middlesex, on 4 March 1923 – of Patrick Moore, the prolific author, broadcaster and accomplished amateur astronomer whose name was associated with the *Yearbook of Astronomy* for almost half a century.

The Monthly Sky Notes and Articles section of the book concludes with a trio of articles penned by Neil Norman, these being *Comets in 2023*, *Minor Planets in 2023* and *Meteor Showers in 2023*, all three titles being fairly self-explanatory describing as they do the occurrence and visibility of examples of these three classes of object during and throughout the year.

In his article *Recent Advances in Astronomy* Rod Hine brings together some more astronomical superlatives, from investigations in cosmic microwave background (CMB) absorption lines at the highest red-shift possible; newly-discovered giant radio galaxies; exciting new observations of fast radio bursts (FRBs); the closest black hole discovered so far; not forgetting the contributions made by "citizen science" projects.

This is followed by *Recent Advances in Solar System Exploration* in which Peter Rea updates us on the progress of a number of planetary missions. Mars is a frequent target for exploration, and during the last two decades we have learnt more about the planet than in the previous two hundred years of telescopic observations. There has never been a more productive time for Martian exploration, as shown in a discussion on the eleven missions currently active on the surface or in orbit around the red planet. Recent discoveries at Jupiter suggest the possibility of water below the surface of some of the large Galilean satellites. The European JUICE mission will go in search for it and should return exciting science from these fascinating worlds.

In his article *Anniversaries in 2023* Neil Haggath commemorates the birth of Nicolaus Copernicus, the man who changed for ever our view of the Solar System. He also notes the births of Ejnar Hertzsprung and Karl Schwarzschild (one day apart), the deaths of the great observer Edward Emerson Barnard and of Gerard Kuiper (of Kuiper Belt fame), as well as celebrating the 50th anniversary of the launch of the Skylab space station.

Marking one of the shoulders of Orion the Hunter, the iconic star Betelgeuse attracted widespread attention when it faded dramatically during the winter of 2019–2020, and some people speculated that Betelgeuse might be about to explode as a supernova. However, the star is now more or less back to normal. In the article *Betelgeuse*, Tracie Heywood questions how unusual the fade was, explains why the fade was so deep and outlines what detailed investigations have revealed regarding the likely future for Betelgeuse.

In the 2022 edition, regular contributor David M. Harland recounted how the Search for Extraterrestrial Intelligence (SETI) began in the radio spectrum in the early 1960s. Now, in his article *Optical SETI at Harvard*, he describes an extension of the search into the optical range by a team led by Paul Horowitz of Harvard University which has developed instruments to scan the sky for pulsed laser signals.

Much has been written about the early Universe. Astronomers can describe its state in detail from the tiniest fraction of a second after the Big Bang. In his surprisingly upbeat article *A Brief History of the End of the Universe*, David Harper looks instead to the very distant future, when the stars will go out, black holes will evaporate, and matter itself ceases to exist. He also considers the fate of our home planet over the next five billion years as the Sun approaches old age.

In his article *The Evolution of "Multi-Pixel" Radio Telescopes* Rod Hine reviews the development of radio telescopes from unwieldy contraptions via huge structures to sophisticated arrays of extreme resolution and precision imaging capability, in just six decades. This is followed by *Mission to Mars: Countdown to Building a Brave New World: The Bare Necessities of Life* by Martin Braddock, the third in a series of articles scheduled to appear in the Yearbook of Astronomy throughout the 2020s and which will keep the reader fully up to date with the ongoing preparations geared towards sending a manned mission to Mars at or around the turn of the decade.

A somewhat baffling and worrying phenomenon, in today's world of technology, is the decline in many people's understanding of science, and the increasing numbers who buy into bizarre antiscientific beliefs. In his article *The Ability to Believe: The Bizarre Worlds of Astronomical Antireality*, Neil Haggath explores the strange beliefs of Flat Earthers, Young Earth Creationists and other deniers of science.

Lynne Marie Stockman checks the interstellar speed limit in *The Astronomers' Stars: Life in the Fast Lane*, part two of a series exploring unusual stars named after the astronomers who discovered them. Barnard's Star may be the best known of the stars living 'life in the fast lane' but proper motion studies from the eighteenth century onwards have revealed a number of other objects passing by at high velocity, including Piazzi's Flying Star (the first star to have its true distance measured), the superflare-prone Argelander's Star, and the intergalactic traveller Kapteyn's Star.

During the first half of the nineteenth century, an energetic schoolteacher from New England made significant contributions to America's burgeoning interest in astronomy. In his article *Elijah Hinsdale Burritt: Geographer of the Heavens* author Richard Sanderson sheds light on the life of a man who wrote one of the best-selling astronomy books and star atlases of its time. Even today, collectors treasure original copies of the handful of books penned by Burritt before his life was tragically cut short.

The future of spaceflight has been the subject of numerous articles over the years. In the past the future always seemed far away, with promises that were more like hopes and dreams. Now however, the future is actually upon us. In his article *The Future of Spaceflight*, Andrew Lound examines the current trend towards regular commercial spaceflight and the opportunities that it will present. For the first time we can safely say that within a generation humanity will be a space faring species, and that some of their grandchildren will be born off the Earth.

In part two of her *Male Mentors for Women in Astronomy* series, Mary McIntyre tells us about the brother-sister partnership of William and Caroline Herschel. We all know that William saved Caroline from her "Cinderella" life, but this article takes a deeper dive into Caroline's story. She was a remarkable woman who triumphed against great adversity to become an amazing astronomer. She was the first woman to be paid a salary for working in science and became a true icon for all women in astronomy.

Unlike most facilities in which science is pursued, an astronomical observatory possesses a cathedral-like quality that evokes a noble probing of mysteries at great distances in space and time. And yet, these temples of science are subject to obsolescence, light pollution, shifting goals, shrinking budgets, and even administrative whimsy that may lead to their closure. In his article *The Closing of Historic Observatories*, Harold A. McAlister examines several such instances when historic observatories have been resurrected from their demise to find new roles in science, education and community enrichment. Unfortunately, there are some unhappy outcomes as well.

The final section of the book starts off with *Some Interesting Variable Stars* by Tracie Heywood which contains useful information on variables as well as predictions for timings of minimum brightness of the famous eclipsing binary Algol for 2023. *Some Interesting Double Stars* and *Some Interesting Nebulae, Star Clusters and Galaxies* present a selection of objects for you to seek out in the night sky. The lists included here are by no means definitive and may well omit your favourite celestial targets. If this is the case, please let us know and we will endeavour to include these in future editions of the *Yearbook of Astronomy*.

Next we have a selection of *Astronomical Organizations*, which lists organizations and associations across the world through which you can further pursue your interest and participation in astronomy (if there are any that we have omitted please let us know) and *Our Contributors*, which contains brief background details of the numerous writers who have contributed to this edition of the Yearbook.

The book rounds off with a recently-introduced feature for the *Yearbook of Astronomy*, this being a selection of advertisements for national and international astronomy- and space-related organizations throughout the world. This is intended to supplement the list of *Astronomical Organizations* (see previous paragraph) and first appeared in the 2022 edition. The selection has been expanded for *Yearbook of Astronomy 2023* and will continue to be a feature of the Yearbook when space permits.

Over time new topics and themes will be introduced into the *Yearbook of Astronomy* to allow it to keep pace with the increasing range of skills, techniques and observing methods now open to amateur astronomers, this in addition to articles relating to our rapidly-expanding knowledge of the Universe in which we live. There will be an interesting mix, some articles written at a level which will appeal to the casual reader and some of what may be loosely described as at a more academic level. The intention is to fully maintain and continually increase the usefulness and relevance of the *Yearbook of Astronomy* to the interests of the readership who are, without doubt, the most important aspect of the Yearbook and the reason it exists in the first place. With this in mind, suggestions from readers for further improvements and additions to the Yearbook content are always welcomed. After all, the book is written for you . . .

As ever, grateful thanks are extended to those individuals who have contributed a great deal of time and effort to the *Yearbook of Astronomy 2023*, including David Harper, who has provided updated versions of his excellent Monthly Star Charts. These were generated specifically for what has been described as the new generation of the *Yearbook of Astronomy*, and the charts add greatly to the overall value of the book to star gazers. Equally important are the efforts of Lynne Marie Stockman who has put together the *Monthly Sky Notes*. Their combined efforts have produced what can justifiably be described as the backbone of the Yearbook of Astronomy. Also worthy of mention is Mat Blurton, who has done an excellent job typesetting the Yearbook. Also Jonathan Wright, Charlotte Mitchell, Lori Jones, Janet Brookes, Paul Wilkinson, Charlie Simpson and Rosie Crofts of Pen & Sword Books Ltd for their efforts in producing and promoting the *Yearbook of Astronomy 2023*, the latest edition of this much-loved and iconic publication.

Brian Jones - Editor
Bradford, West Riding of Yorkshire
October 2021

Preface

The information given in this edition of the Yearbook of Astronomy is in narrative form. The positions of the planets given in the Monthly Sky Notes often refer to the constellations in which they lie at the time. These can be found on the star charts which collectively show the whole sky via two charts depicting the northern and southern circumpolar stars and forty-eight charts depicting the main stars and constellations for each month of the year. The northern and southern circumpolar charts show the stars that are within 45° of the two celestial poles, while the monthly charts depict the stars and constellations that are visible throughout the year from Europe and North America or from Australia and New Zealand. The monthly charts overlap the circumpolar charts. Wherever you are on the Earth, you will be able to locate and identify the stars depicted on the appropriate areas of the chart(s).

There are numerous star atlases available that offer more detailed information, such as *Sky & Telescope's POCKET SKY ATLAS* and *Norton's STAR ATLAS and Reference Handbook* to name but a couple. In addition, more precise information relating to planetary positions and so on can be found in a number of publications, a good example of which is *The Handbook of the British Astronomical Association*, as well as many of the popular astronomy magazines such as the British monthly periodicals *Sky at Night* and *Astronomy Now* and the American monthly magazines *Astronomy* and *Sky & Telescope*.

About Time

Before the late eighteenth century, the biggest problem affecting mariners sailing the seas was finding their position. Latitude was easily determined by observing the altitude of the pole star above the northern horizon. Longitude, however, was far more difficult to measure. The inability of mariners to determine their longitude often led to them getting lost, and on many occasions shipwrecked. To address this problem King Charles II established the Royal Observatory at Greenwich in 1675 and from here, Astronomers Royal began the process of measuring and cataloguing the stars as they passed due south across the Greenwich meridian.

Now mariners only needed an accurate timepiece (the chronometer invented by Yorkshire-born clockmaker John Harrison) to display GMT (Greenwich Mean Time). Working out the local standard time onboard ship and subtracting this from GMT gave the ship's longitude (west or east) from the Greenwich meridian. Therefore mariners always knew where they were at sea and the longitude problem was solved.

Astronomers use a time scale called Universal Time (UT). This is equivalent to Greenwich Mean Time and is defined by the rotation of the Earth. The Yearbook of Astronomy gives all times in UT rather than in the local time for a particular city or country. Times are expressed using the 24-hour clock, with the day beginning at midnight, denoted by 00:00. Universal Time (UT) is related to local mean time by the formula:

Local Mean Time = UT – west longitude

In practice, small differences in longitude are ignored and the observer will use local clock time which will be the appropriate Standard (or Zone) Time. As the formula indicates, places in west longitude will have a Standard Time slow on UT, while those in east longitude will have a Standard Time fast on UT. As examples we have:

Standard Time in

New Zealand	UT +12 hours
Victoria, NSW	UT +10 hours
Western Australia	UT + 8 hours
South Africa	UT + 2 hours
British Isles	UT
Newfoundland Standard Time	UT −3 hours 30 minutes
Atlantic Standard Time	UT −4 hours
Eastern Standard Time	UT −5 hours
Central Standard Time	UT −6 hours
Mountain Standard Time	UT −7 hours
Pacific Standard Time	UT −8 hours
Alaska Standard Time	UT −9 hours
Hawaii-Aleutian Standard Time	UT −10 hours

During the periods when Summer Time (also called Daylight Saving Time) is in use, one hour must be added to Standard Time to obtain the appropriate Summer/ Daylight Saving Time. For example, Pacific Daylight Time is UT −7 hours.

Using the Yearbook of Astronomy
as an Observing Guide

Notes on the Monthly Star Charts

The star charts on the following pages show the night sky throughout the year. There are two sets of charts, one for use by observers in the Northern Hemisphere and one for those in the Southern Hemisphere. The first set is drawn for latitude 52°N and can be used by observers in Europe, Canada and most of the United States. The second set is drawn for latitude 35°S and show the stars as seen from Australia and New Zealand. Twelve pairs of charts are provided for each of these latitudes.

Each pair of charts shows the entire sky as two semi-circular half-sky views, one looking north and the other looking south. A given pair of charts can be used at different times of year. For example, chart 1 shows the night sky at midnight on 21 December, but also at 2am on 21 January, 4am on 21 February and so forth. The accompanying table will enable you to select the correct chart for a given month and time of night. The caption next to each chart also lists the dates and times of night for which it is valid.

The charts are intended to help you find the more prominent constellations and other objects of interest mentioned in the monthly observing notes. To avoid the charts becoming too crowded, only stars of magnitude 4.5 or brighter are shown. This corresponds to stars that are bright enough to be seen from any dark suburban garden on a night when the Moon is not too close to full phase.

Each constellation is depicted by joining selected stars with lines to form a pattern. There is no official standard for these patterns, so you may occasionally find different patterns used in other popular astronomy books for some of the constellations.

Any map projection from a sphere onto a flat page will by necessity contain some distortions. This is true of star charts as well as maps of the Earth. The distortion on the half-sky charts is greatest near the semi-circular boundary of each chart, where it may appear to stretch constellation patterns out of shape.

The charts also show selected deep-sky objects such as galaxies, nebulae and star clusters. Many of these objects are too faint to be seen with the naked eye, and you will need binoculars or a telescope to observe them. Please refer to the table of deep-sky objects for more information.

Planetary Apparition Diagrams

The diagrams of the apparitions of Mercury and Venus show the position of the respective planet in the sky at the moment of sunrise or sunset throughout the entire apparition. Two sets of positions are plotted on each chart: for latitude 52° North (blue line) and for latitude 35° South (red line). A thin dotted line denotes the portion of the apparition which falls outside the year covered by this edition of the Yearbook. A white dot indicates the position of Venus on the first day of each month, or of Mercury on the first, eleventh and 21st of the month. The day of greatest elongation (GE) is also marked by a white dot. Note that the dots do NOT indicate the magnitude of the planet.

The finder chart for Mars shows its path during the first eight months of 2023, when it is an evening object as it moves away from opposition in December of last year. Mars traverses more than 100° in ecliptic longitude during this period, moving from Taurus in January to Virgo in August, so the chart is based upon the ecliptic, which runs across the centre of the chart from right to left. The position of Mars is indicated on the 1st of each month, and at its most westerly on 12 January, when it resumes direct motion. Stars are shown to magnitude 5.5. Note that the sizes of the Mars dots do NOT indicate its magnitude.

The finder charts for Jupiter, Saturn, Uranus and Neptune show the paths of the planets throughout the year. The position of each planet is indicated at opposition and at stationary points, as well as the start and end of the year and on the 1st of each month (1st of April, July and October only for Uranus and Neptune) where these dates do not fall too close to an event that is already marked. Stars are shown to magnitude 5.5 on the charts for Jupiter and Saturn. On the Uranus chart, stars are shown to magnitude 8; on the Neptune chart, the limiting magnitude is 10. In both cases, this is approximately two magnitudes fainter than the planet itself. Right Ascension and Declination scales are shown for the epoch J2000 to allow comparison with modern star charts. Note that the sizes of the dots denoting the planets do NOT indicate their magnitudes.

Selecting the Correct Charts

The table below shows which of the charts to use for particular dates and times throughout the year and will help you to select the correct pair of half-sky charts for any combination of month and time of night.

The Earth takes 23 hours 56 minutes (and 4 seconds) to rotate once around its axis with respect to the fixed stars. Because this is around four minutes shorter than a full 24 hours, the stars appear to rise and set about 4 minutes earlier on each successive day, or around an hour earlier each fortnight. Therefore, as well as showing the stars at 10pm (22h in 24-hour notation) on 21 January, chart 1 also depicts the sky at 9pm (21h) on 6 February, 8pm (20h) on 21 February and 7pm (19h) on 6 March.

The times listed do not include summer time (daylight saving time), so if summer time is in force you must subtract one hour to obtain standard time (GMT if you are in the United Kingdom) before referring to the chart. For example, to find the correct chart for mid-September in the northern hemisphere at 3am summer time, first of all subtract one hour to obtain 2am (2h) standard time. Then you can consult the table, where you will find that you should use chart 11.

The table does not indicate sunrise, sunset or twilight. In northern temperate latitudes, the sky is still light at 18h and 6h from April to September, and still light at 20h and 4h from May to August. In Australia and New Zealand, the sky is still light at 18h and 6h from October to March, and in twilight (with only bright stars visible) at 20h and 04h from November to January.

Local Time	18h	20h	22h	0h	2h	4h	6h
January	11	12	1	2	3	4	5
February	12	1	2	3	4	5	6
March	1	2	3	4	5	6	7
April	2	3	4	5	6	7	8
May	3	4	5	6	7	8	9
June	4	5	6	7	8	9	10
July	5	6	7	8	9	10	11
August	6	7	8	9	10	11	12
September	7	8	9	10	11	12	1
October	8	9	10	11	12	1	2
November	9	10	11	12	1	2	3
December	10	11	12	1	2	3	4

Legend to the Star Charts

STARS		DEEP-SKY OBJECTS	
Symbol	Magnitude	Symbol	Type of object
•	0 or brighter	✳	Open star cluster
•	1	○	Globular star cluster
•	2	□	Nebula
•	3	▦	Cluster with nebula
·	4	○	Planetary nebula
·	5	◌	Galaxy
✦	Double star		Magellanic Clouds
◉	Variable star		

Star Names

There are over 200 stars with proper names, most of which are of Roman, Greek or Arabic origin although only a couple of dozen or so of these names are used regularly. Examples include Arcturus in Boötes, Castor and Pollux in Gemini and Rigel in Orion.

A system whereby Greek letters were assigned to stars was introduced by the German astronomer and celestial cartographer Johann Bayer in his star atlas Uranometria, published in 1603. Bayer's system is applied to the brighter stars within any particular constellation, which are given a letter from the Greek alphabet followed by the genitive case of the constellation in which the star is located. This genitive case is simply the Latin form meaning 'of' the constellation. Examples are the stars Alpha Boötis and Beta Centauri which translate literally as 'Alpha of Boötes' and 'Beta of the Centaur'.

As a general rule, the brightest star in a constellation is labelled Alpha (α), the second brightest Beta (β), and the third brightest Gamma (γ) and so on, although there are some constellations where the system falls down. An example is Gemini where the principal star (Pollux) is designated Beta Geminorum, the second brightest (Castor) being known as Alpha Geminorum.

There are only 24 letters in the Greek alphabet, the consequence of which was that the fainter naked eye stars needed an alternative system of classification. The system in popular use is that devised by the first Astronomer Royal John Flamsteed in which the stars in each constellation are listed numerically in order from west to east. Although many of the brighter stars within any particular constellation will have both Greek letters and Flamsteed numbers, the latter are generally used only when a star does not have a Greek letter.

The Greek Alphabet

α	Alpha	ι	Iota	ρ	Rho
β	Beta	κ	Kappa	σ	Sigma
γ	Gamma	λ	Lambda	τ	Tau
δ	Delta	μ	Mu	υ	Upsilon
ε	Epsilon	ν	Nu	φ	Phi
ζ	Zeta	ξ	Xi	χ	Chi
η	Eta	ο	Omicron	ψ	Psi
θ	Theta	π	Pi	ω	Omega

The Names of the Constellations

On clear, dark, moonless nights, the sky seems to teem with stars although in reality you can never see more than a couple of thousand or so at any one time when looking with the unaided eye. Each and every one of these stars belongs to a particular constellation, although the constellations that we see in the sky, and which grace the pages of star atlases, are nothing more than chance alignments. The stars that make up the constellations are often situated at vastly differing distances from us and only appear close to each other, and form the patterns that we see, because they lie in more or less the same direction as each other as seen from Earth.

A large number of the constellations are named after mythological characters, and were given their names thousands of years ago. However, those star groups lying close to the south celestial pole were discovered by Europeans only during the last few centuries, many of these by explorers and astronomers who mapped the stars during their journeys to lands under southern skies. This resulted in many of the newer constellations having modern-sounding names, such as Octans (the Octant) and Microscopium (the Microscope), both of which were devised by the French astronomer Nicolas Louis De La Caille during the early 1750s.

Over the centuries, many different suggestions for new constellations have been put forward by astronomers who, for one reason or another, felt the need to add new groupings to star charts and to fill gaps between the traditional constellations. Astronomers drew up their own charts of the sky, incorporating their new groups into them. A number of these new constellations had cumbersome names, notable examples including Officina Typographica (the Printing Shop) introduced by the German astronomer Johann Bode in 1801; Sceptrum Brandenburgicum (the Sceptre of Brandenburg) introduced by the German astronomer Gottfried Kirch in 1688; Taurus Poniatovii (Poniatowski's Bull) introduced by the Polish-Lithuanian astronomer Martin Odlanicky Poczobut in 1777; and Quadrans Muralis (the Mural Quadrant) devised by the French astronomer Joseph-Jerôme de Lalande in1795. Although these have long since been rejected, the latter has been immortalised by the annual Quadrantid meteor shower, the radiant of which lies in an area of sky formerly occupied by Quadrans Muralis.

During the 1920s the International Astronomical Union (IAU) systemised matters by adopting an official list of 88 accepted constellations, each with official spellings and abbreviations. Precise boundaries for each constellation were then drawn up so that every point in the sky belonged to a particular constellation.

The abbreviations devised by the IAU each have three letters which in the majority of cases are the first three letters of the constellation name, such as AND for Andromeda, EQU for Equuleus, HER for Hercules, ORI for Orion and so on. This trend is not strictly adhered to in cases where confusion may arise. This happens with the two constellations Leo (abbreviated LEO) and Leo Minor (abbreviated LMI). Similarly, because Triangulum (TRI) may be mistaken for Triangulum Australe, the latter is abbreviated TRA. Other instances occur with Sagitta (SGE) and Sagittarius (SGR) and with Canis Major (CMA) and Canis Minor (CMI) where the first two letters from the second names of the constellations are used. This is also the case with Corona Australis (CRA) and Corona Borealis (CRB) where the first letter of the second name of each constellation is incorporated. Finally, mention must be made of Crater (CRT) which has been abbreviated in such a way as to avoid confusion with the aforementioned CRA (Corona Australis).

The table shown on the following pages contains the name of each of the 88 constellations together with the translation and abbreviation of the constellation name. The constellations depicted on the monthly star charts are identified with their abbreviations rather than the full constellation names.

The Constellations

Andromeda	Andromeda	AND
Antlia	The Air Pump	ANT
Apus	The Bird of Paradise	APS
Aquarius	The Water Carrier	AQR
Aquila	The Eagle	AQL
Ara	The Altar	ARA
Aries	The Ram	ARI
Auriga	The Charioteer	AUR
Boötes	The Herdsman	BOO
Caelum	The Graving Tool	CAE
Camelopardalis	The Giraffe	CAM
Cancer	The Crab	CNC
Canes Venatici	The Hunting Dogs	CVN
Canis Major	The Great Dog	CMA
Canis Minor	The Little Dog	CMI
Capricornus	The Goat	CAP
Carina	The Keel	CAR
Cassiopeia	Cassiopeia	CAS
Centaurus	The Centaur	CEN
Cepheus	Cepheus	CEP
Cetus	The Whale	CET
Chamaeleon	The Chameleon	CHA
Circinus	The Pair of Compasses	CIR
Columba	The Dove	COL
Coma Berenices	Berenice's Hair	COM
Corona Australis	The Southern Crown	CRA
Corona Borealis	The Northern Crown	CRB
Corvus	The Crow	CRV
Crater	The Cup	CRT
Crux	The Cross	CRU
Cygnus	The Swan	CYG

Delphinus	The Dolphin	DEL
Dorado	The Goldfish	DOR
Draco	The Dragon	DRA
Equuleus	The Foal	EQU
Eridanus	The River	ERI
Fornax	The Furnace	FOR
Gemini	The Twins	GEM
Grus	The Crane	GRU
Hercules	Hercules	HER
Horologium	The Pendulum Clock	HOR
Hydra	The Water Snake	HYA
Hydrus	The Lesser Water Snake	HYI
Indus	The Indian	IND
Lacerta	The Lizard	LAC
Leo	The Lion	LEO
Leo Minor	The Lesser Lion	LMI
Lepus	The Hare	LEP
Libra	The Scales	LIB
Lupus	The Wolf	LUP
Lynx	The Lynx	LYN
Lyra	The Lyre	LYR
Mensa	The Table Mountain	MEN
Microscopium	The Microscope	MIC
Monoceros	The Unicorn	MON
Musca	The Fly	MUS
Norma	The Level	NOR
Octans	The Octant	OCT
Ophiuchus	The Serpent Bearer	OPH
Orion	Orion	ORI
Pavo	The Peacock	PAV
Pegasus	Pegasus	PEG
Perseus	Perseus	PER

Phoenix	The Phoenix	PHE
Pictor	The Painter's Easel	PIC
Pisces	The Fish	PSC
Piscis Austrinus	The Southern Fish	PSA
Puppis	The Stern	PUP
Pyxis	The Mariner's Compass	PYX
Reticulum	The Net	RET
Sagitta	The Arrow	SGE
Sagittarius	The Archer	SGR
Scorpius	The Scorpion	SCO
Sculptor	The Sculptor	SCL
Scutum	The Shield	SCT
Serpens Caput and Cauda	The Serpent	SER

Sextans	The Sextant	SEX
Taurus	The Bull	TAU
Telescopium	The Telescope	TEL
Triangulum	The Triangle	TRI
Triangulum Australe	The Southern Triangle	TRA
Tucana	The Toucan	TUC
Ursa Major	The Great Bear	UMA
Ursa Minor	The Little Bear	UMI
Vela	The Sail	VEL
Virgo	The Virgin	VIR
Volans	The Flying Fish	VOL
Vulpecula	The Fox	VUL

The Monthly Star Charts

Northern Hemisphere Star Charts

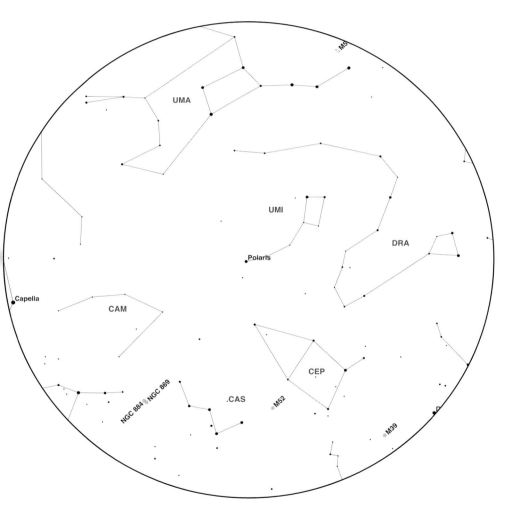

This chart shows stars lying at declinations between +45 and +90 degrees. These constellations are circumpolar for observers in Europe and North America.

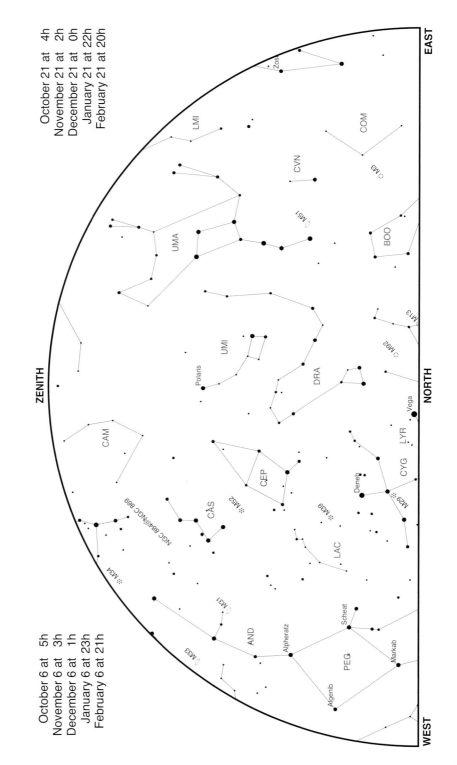

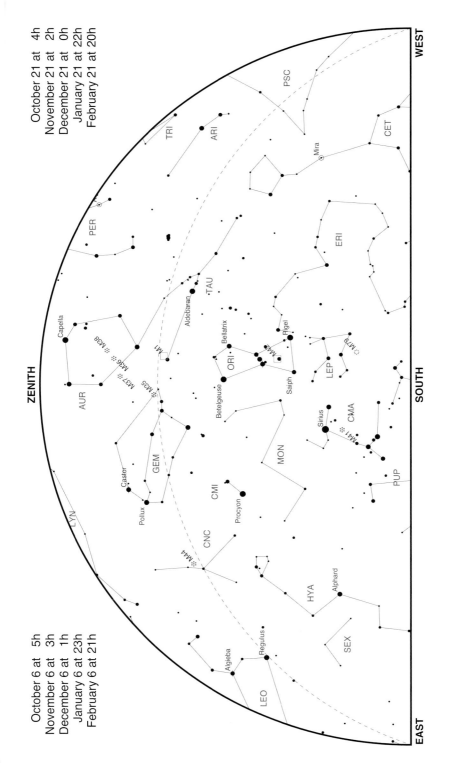

1S

WEST

ZENITH

EAST

SOUTH

October 21 at 4h
November 21 at 2h
December 21 at 0h
January 21 at 22h
February 21 at 20h

October 6 at 5h
November 6 at 3h
December 6 at 1h
January 6 at 23h
February 6 at 21h

PSC

CET

Mira

TRI

ARI

PER

ERI

Capella

TAU

Aldebaran

M38
M36
M37
M35

M1

Bellatrix

Rigel

ORI

M42

Saiph

M79

LEP

AUR

Betelgeuse

Sirius

CMA

M41

Castor

GEM

MON

PUP

Pollux

CMI

Procyon

LYN

CNC

M44

HYA

Alphard

Algieba

Regulus

SEX

LEO

2N

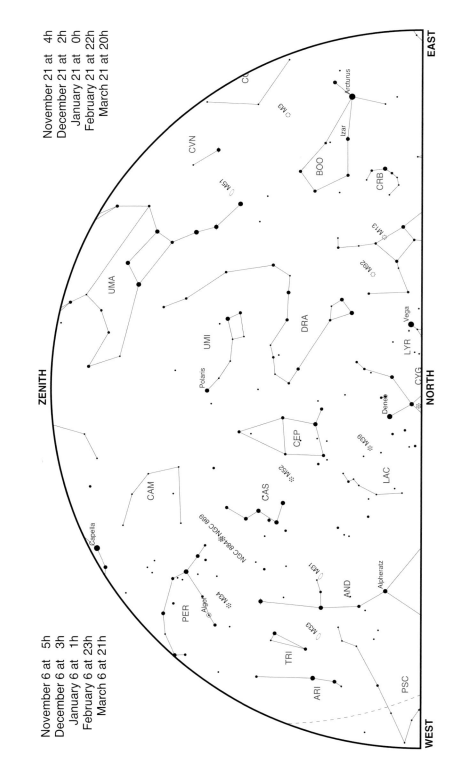

November 21 at 4h
December 21 at 2h
January 21 at 0h
February 21 at 22h
March 21 at 20h

November 6 at 5h
December 6 at 3h
January 6 at 1h
February 6 at 23h
March 6 at 21h

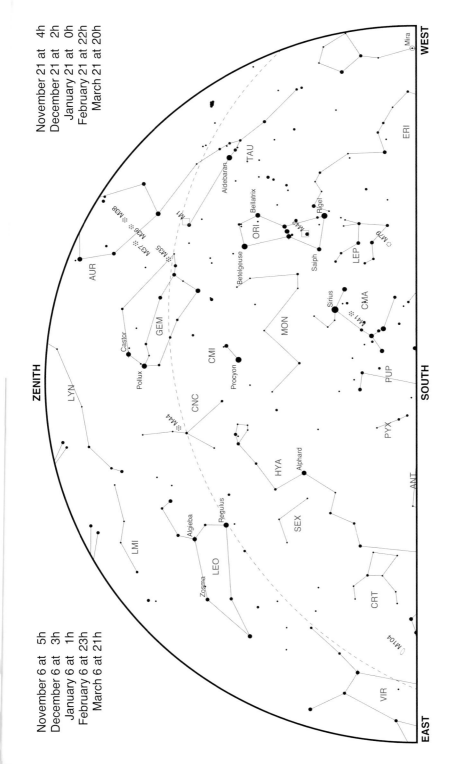

2S

WEST

ZENITH

EAST

SOUTH

Mira
ERI
TAU
Aldebaran
Bellatrix
ORI
Rigel
M42
Betelgeuse
Saiph
LEP
M79
Sirius
CMA
M41
M35
M37
M36
M38
AUR
M1
Castor
GEM
Pollux
CMI
MON
Procyon
PUP
PYX
CNC
M44
LYN
HYA
Alphard
ANT
SEX
LMI
Algieba
Regulus
LEO
Zosma
CRT
M104
VIR

3N

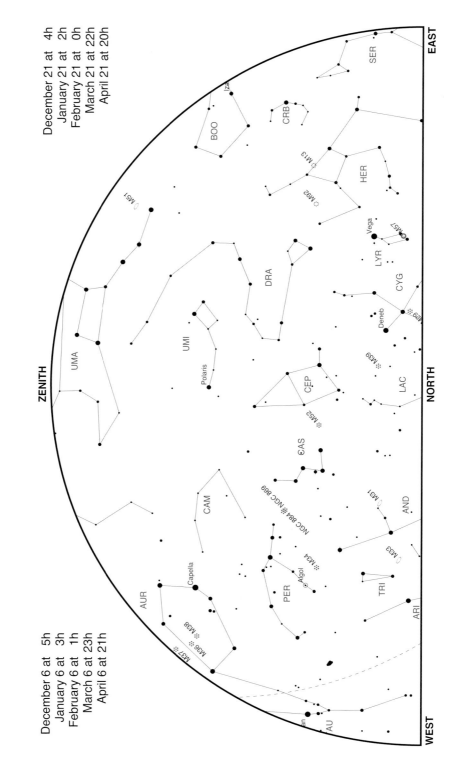

December 21 at 4h
January 21 at 2h
February 21 at 0h
March 21 at 22h
April 21 at 20h

December 6 at 5h
January 6 at 3h
February 6 at 1h
March 6 at 23h
April 6 at 21h

EAST

ZENITH

NORTH

WEST

SER

CRB

Izar

BOO

M13

HER

M92

M57

Vega

LYR

M51

CYG

DRA

B2N

Deneb

UMI

M39

Polaris

LAC

UMA

CEP

M52

CAS

CAM

NGC 884 NGC 869

M31

M34

AND

Capella

Algol

M33

PER

TRI

AUR

ARI

M38

M36

M37

AU

an

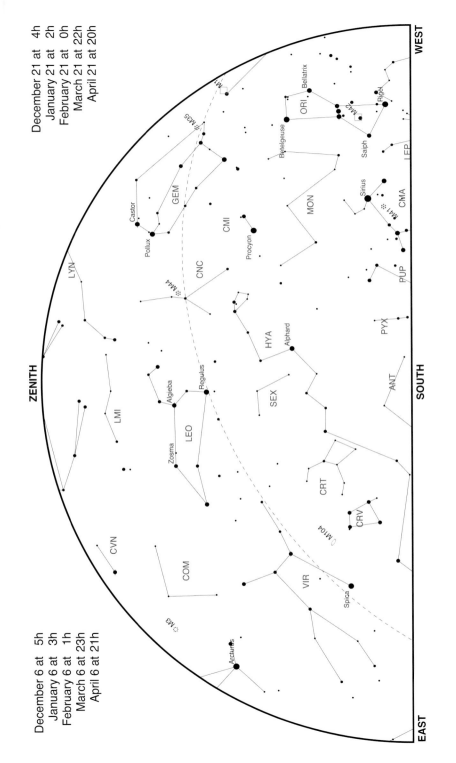

3S

WEST

December 21 at 4h
January 21 at 2h
February 21 at 0h
March 21 at 22h
April 21 at 20h

ZENITH

EAST

December 6 at 5h
January 6 at 3h
February 6 at 1h
March 6 at 23h
April 6 at 21h

SOUTH

ORI
Bellatrix
Rigel
M43
Saiph
Betelgeuse
ECL

Sirius
CMA
M41
CMi
Procyon

MON
PUP

GEM
Pollux
Castor
M35

CNC
M44

LYN

HYA
Alphard

PYX

LEO
Regulus
Algieba
Zosma

SEX

ANT

LMI

CVN

COM

CRT
CRV
M104

VIR
Spica

M3
Arcturus

4N

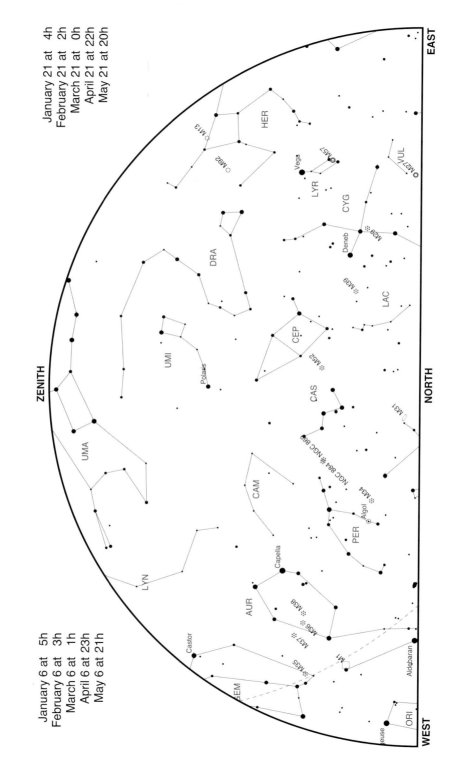

January 6 at 5h
February 6 at 3h
March 6 at 1h
April 6 at 23h
May 6 at 21h

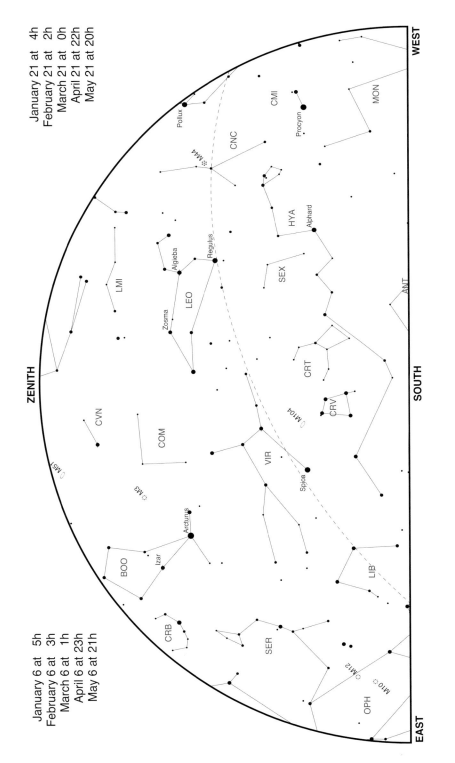

January 21 at 4h
February 21 at 2h
March 21 at 0h
April 21 at 22h
May 21 at 20h

January 6 at 5h
February 6 at 3h
March 6 at 1h
April 6 at 23h
May 6 at 21h

WEST

EAST

SOUTH

ZENITH

MON
Procyon
CMI
CNC
M44
Pollux
Regulus
Algieba
LEO
Zosma
LMI
HYA
Alphard
SEX
ANT
CRT
M104
CRV
VIR
Spica
LIB
CVN
M51
COM
M3
Arcturus
Izar
BOO
CRB
SER
M5
M10
M12
OPH

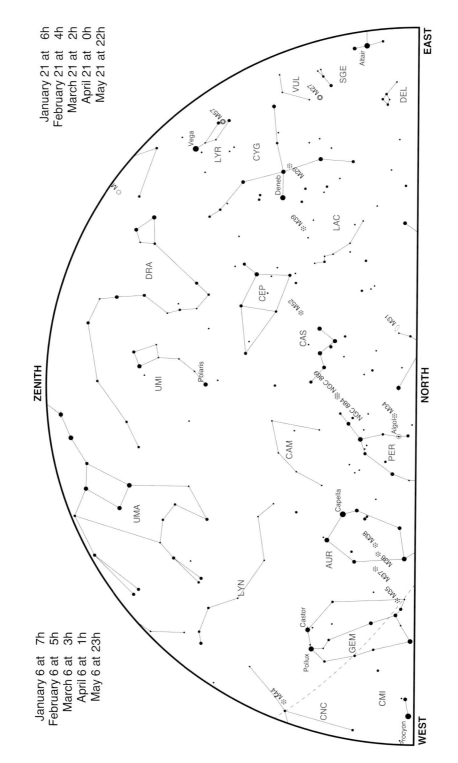

5N

January 6 at 7h
February 6 at 5h
March 6 at 3h
April 6 at 1h
May 6 at 23h

EAST

ZENITH

NORTH

WEST

Altair
SGE
VUL
M27
DEL
Vega
LYR
M57
CYG
M29
Deneb
M39
LAC
DRA
CEP
M52
CAS
M31
UMI
Polaris
NGC 869
NGC 884
M34
Algol
PER
CAM
AUR
Capella
M38
M36
M37
M35
UMA
LYN
Castor
Pollux
GEM
Procyon
CMI
M44
CNC

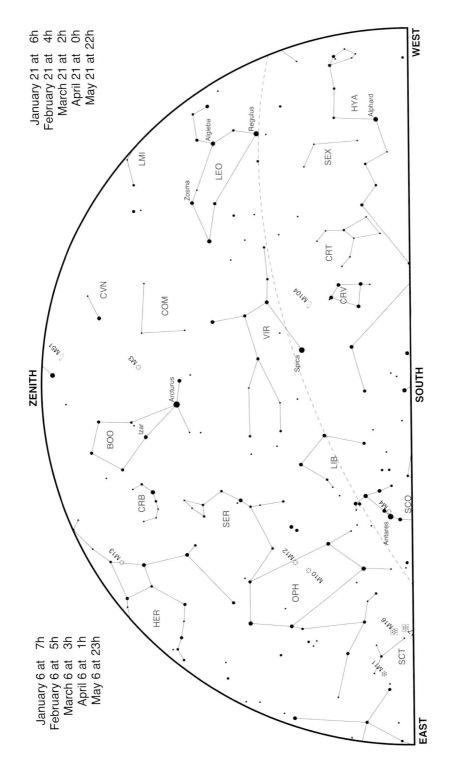

5S

WEST

January 21 at 6h
February 21 at 4h
March 21 at 2h
April 21 at 0h
May 21 at 22h

ZENITH

SOUTH

January 6 at 7h
February 6 at 5h
March 6 at 3h
April 6 at 1h
May 6 at 23h

EAST

LMI

LEO

Algieba

Regulus

Zosma

HYA

Alphard

SEX

CRT

CVN

COM

CRV

M104

VIR

Spica

M3

M51

Arcturus

Izar

BOO

LIB

M4

SCO

Antares

CRB

SER

OPH

M12

M10

HER

M13

M16

M11

SCT

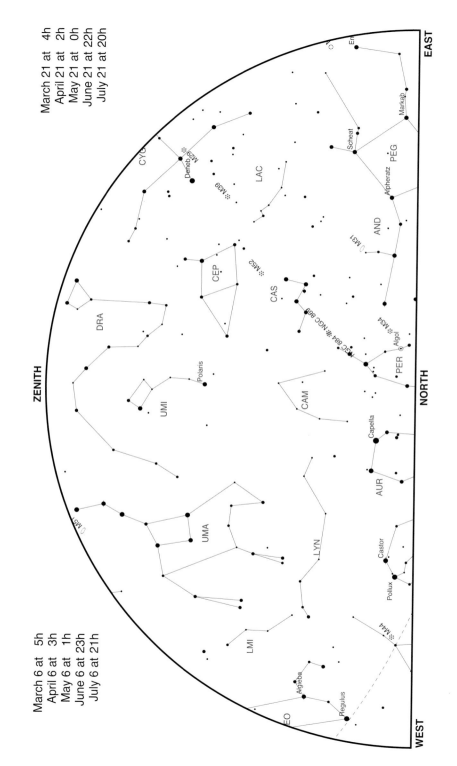

March 21 at 4h
April 21 at 2h
May 21 at 0h
June 21 at 22h
July 21 at 20h

March 6 at 5h
April 6 at 3h
May 6 at 1h
June 6 at 23h
July 6 at 21h

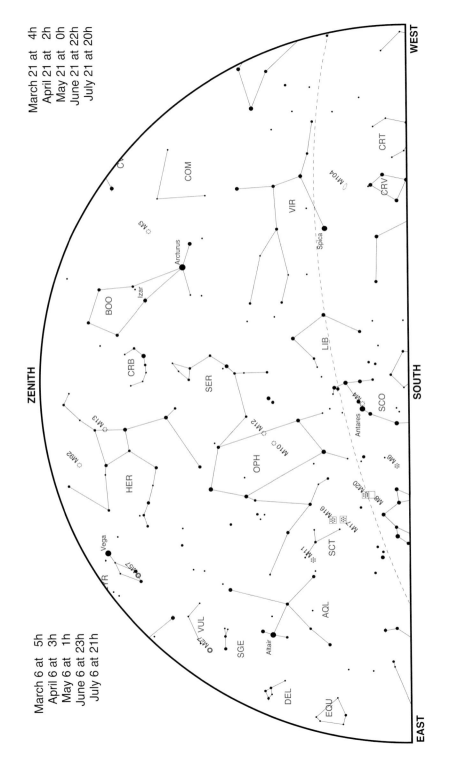

6S

March 21 at 4h
April 21 at 2h
May 21 at 0h
June 21 at 22h
July 21 at 20h

March 6 at 5h
April 6 at 3h
May 6 at 1h
June 6 at 23h
July 6 at 21h

ZENITH

EAST

SOUTH

CRT
CRV
M104
VIR
Spica
COM
M3
Arcturus
BOO
Izar
LIB
CRB
SER
M12
SCO
Antares
M4
M10
OPH
M6
M13
M92
HER
M20
M8
M16
M17
SCT
M11
M57
Vega
VUL
M27
SGE
AQL
Altair
DEL
EQU

7N

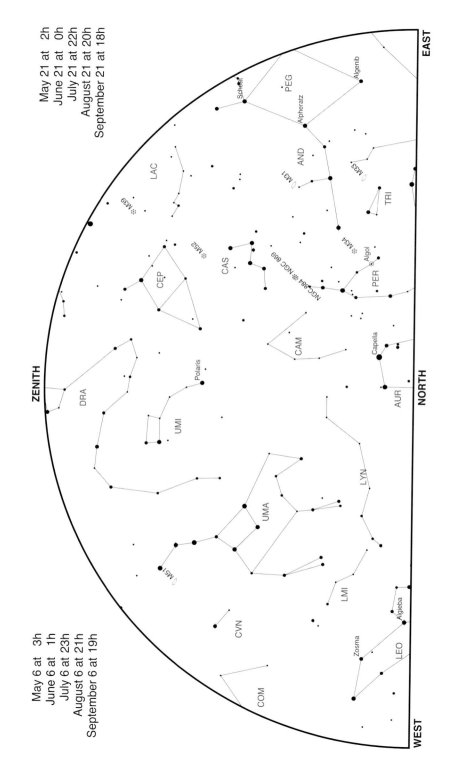

May 21 at 2h
June 21 at 0h
July 21 at 22h
August 21 at 20h
September 21 at 18h

May 6 at 3h
June 6 at 1h
July 6 at 23h
August 6 at 21h
September 6 at 19h

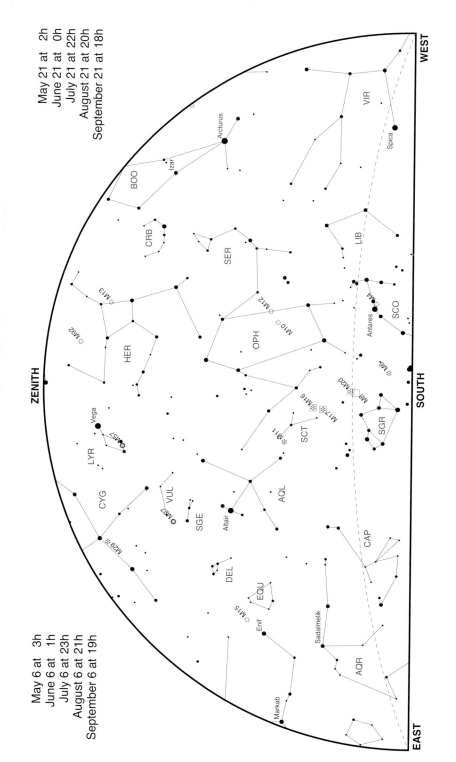

7S

May 21 at 2h
June 21 at 0h
July 21 at 22h
August 21 at 20h
September 21 at 18h

May 6 at 3h
June 6 at 1h
July 6 at 23h
August 6 at 21h
September 6 at 19h

8N

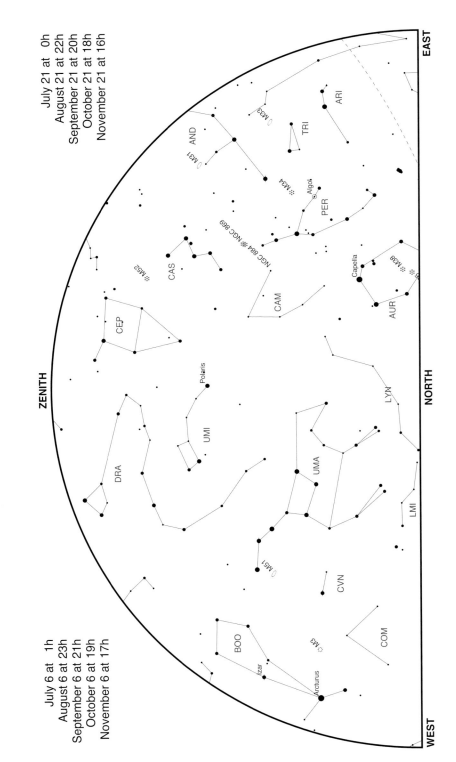

July 21 at 0h
August 21 at 22h
September 21 at 20h
October 21 at 18h
November 21 at 16h

July 6 at 1h
August 6 at 23h
September 6 at 21h
October 6 at 19h
November 6 at 17h

EAST

ZENITH

NORTH

WEST

AND
M31
M34
ARI
TRI
M33
Algol
PER
NGC 884 NGC 869
CAS
M52
CAM
Capella
M38
M36
AUR
CEP
Polaris
UMI
LYN
DRA
UMA
LMI
M51
CVN
BOO
Izar
M3
COM
Arcturus

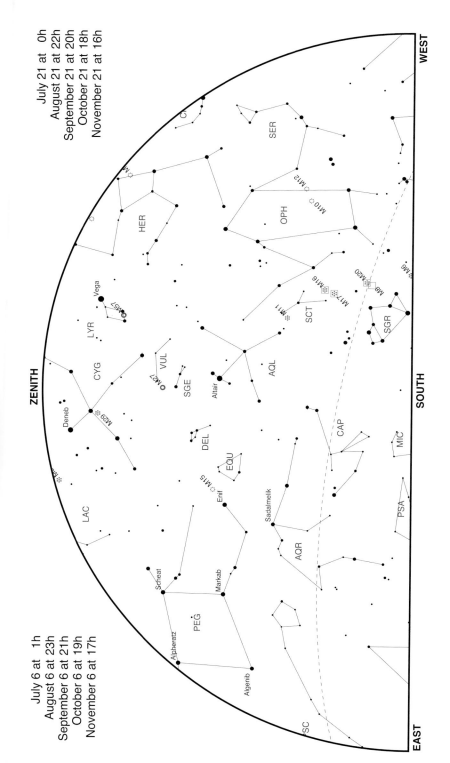

WEST

8S

SER

OPH

M12

M10

Vega

M17 M16

M20

M8

M6

LYR

SCT

SGR

M57

CYG

M11

VUL

Deneb

M29

SGE

AQL

Altair

M27

ZENITH

DEL

EQU

CAP

SOUTH

Enif

M15

MIC

LAC

Sadalmelik

PSA

Scheat

AQR

Markab

PEG

Alpheratz

Algenib

SC

EAST

HER

9N

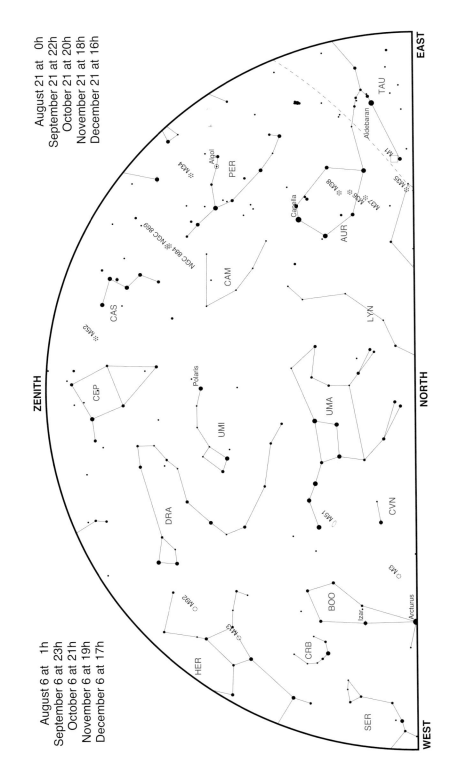

August 21 at 0h
September 21 at 22h
October 21 at 20h
November 21 at 18h
December 21 at 16h

August 6 at 1h
September 6 at 23h
October 6 at 21h
November 6 at 19h
December 6 at 17h

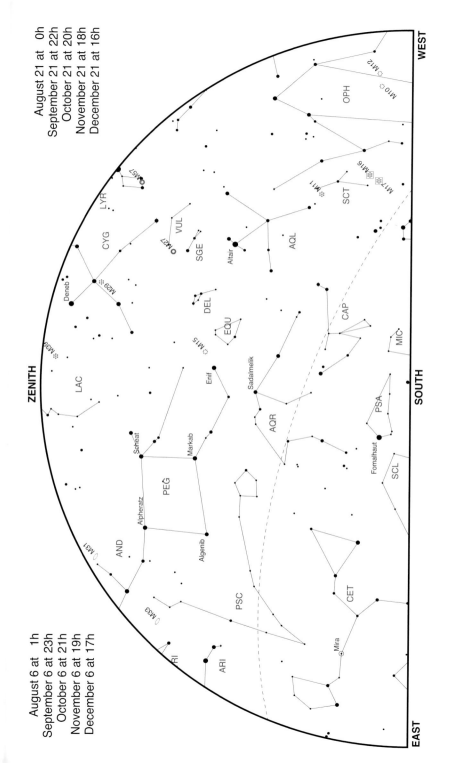

9S

August 21 at 0h
September 21 at 22h
October 21 at 20h
November 21 at 18h
December 21 at 16h

August 6 at 1h
September 6 at 23h
October 6 at 21h
November 6 at 19h
December 6 at 17h

WEST

EAST

SOUTH

ZENITH

M12
M10
OPH
M57
LYR
M17 M16
M11
SCT
CYG
VUL
M27
SGE
Deneb
M29
Altair
AQL
DEL
EQU
CAP
M39
M15
MIC
LAC
Enif
Sadalmelik
PSA
Scheat
AQR
Fomalhaut
Markab
SCL
PEG
Alpheratz
Algenib
AND
M31
CET
M33
PSC
Mira
ARI

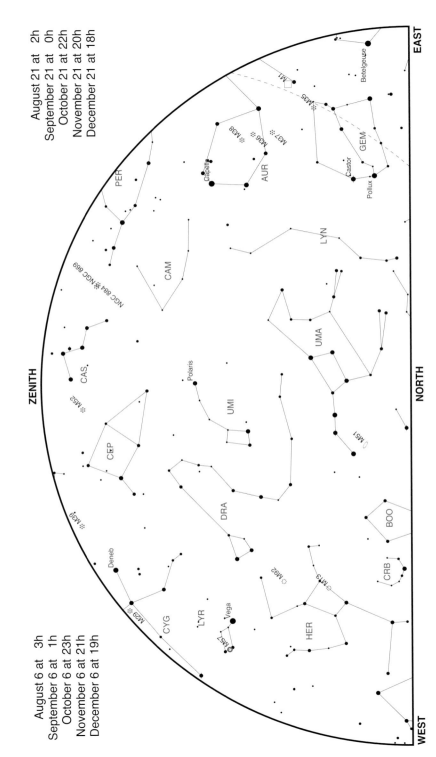

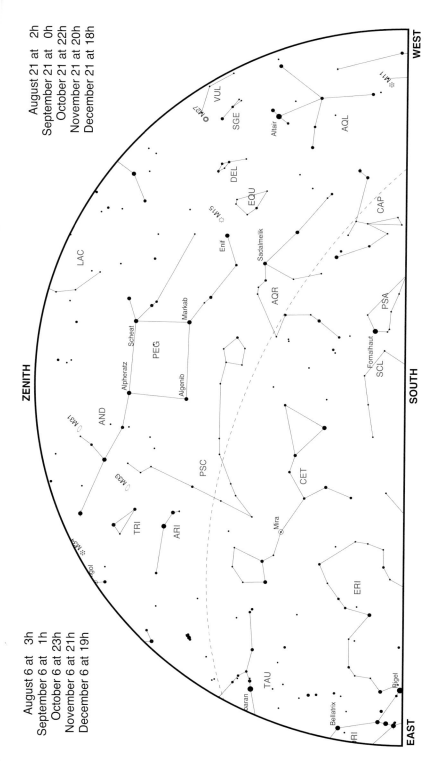

10S

WEST

ZENITH

SOUTH

EAST

VUL
M27
SGE
Altair
AQL
DEL
EQU
M15
Enif
Sadalmelik
CAP
LAC
Scheat
Markab
PEG
Alpheratz
Algenib
AND
M31
AQR
PSA
Fomalhaut
SCL
M33
TRI
ARI
PSC
CET
Mira
Algol
M45
ERI
TAU
Aldebaran
Bellatrix
Rigel
ORI

11N

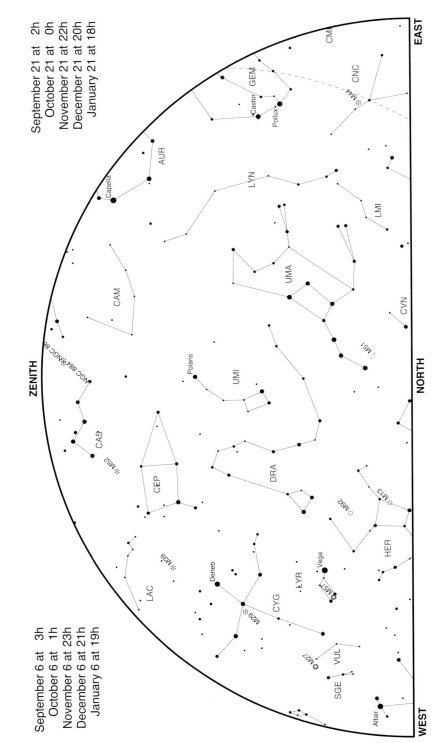

EAST

ZENITH

NORTH

WEST

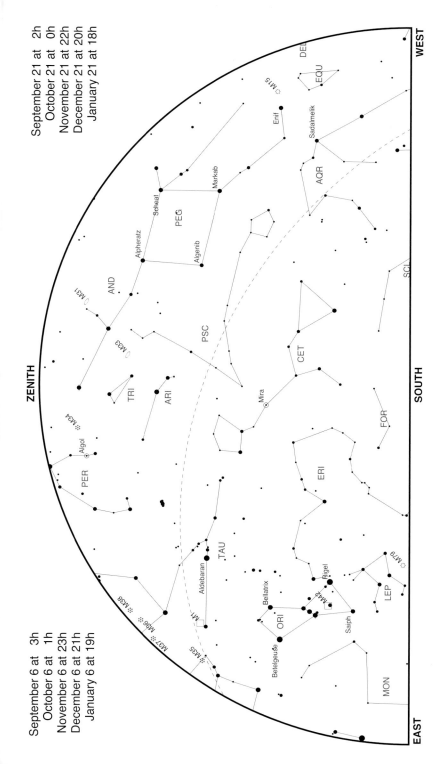

WEST

ZENITH

EAST

SOUTH

September 21 at 2h
October 21 at 0h
November 21 at 22h
December 21 at 20h
January 21 at 18h

September 6 at 3h
October 6 at 1h
November 6 at 23h
December 6 at 21h
January 6 at 19h

DEL
EQU
M15
Enif
Sadalmelik
AQR
Markab
Scheat
PEG
Alpheratz
Algenib
AND
M31
PSC
M33
SCL
CET
TRI
ARI
Mira
FOR
PER
Algol
M34
ERI
M38
M36
TAU
M37
Aldebaran
M1
M35
Bellatrix
Rigel
ORI
M42
Saiph
LEP
M79
Betelgeuse
MON

12N

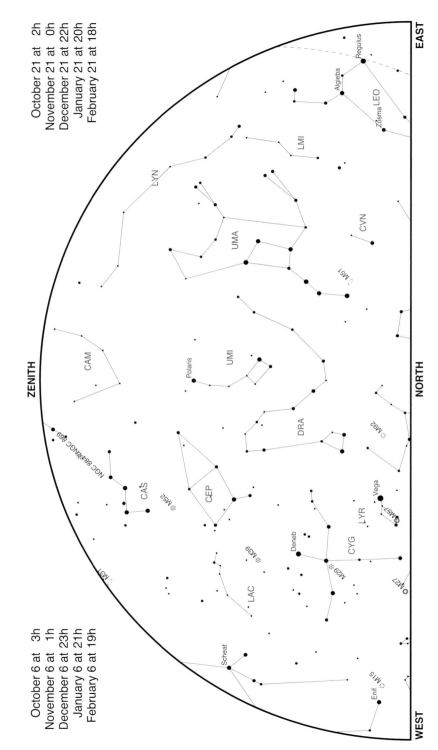

October 21 at 2h
November 21 at 0h
December 21 at 22h
January 21 at 20h
February 21 at 18h

October 6 at 3h
November 6 at 1h
December 6 at 23h
January 6 at 21h
February 6 at 19h

EAST

NORTH

WEST

ZENITH

Regulus
Algieba
Zosma LEO
LMI
LYN
CVN
UMA
M51
CAM
Polaris
UMI
ZENITH
DRA
M92
NGC 884 NGC 869
CAS
M52
CEP
Vega
LYR
M57
Deneb
CYG
M29
M39
M31
LAC
M27
Scheat
M15
Enif

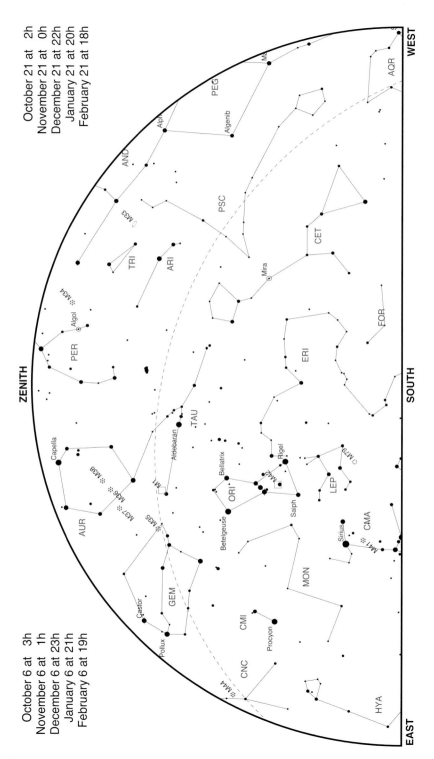

12S

October 21 at 2h
November 21 at 0h
December 21 at 22h
January 21 at 20h
February 21 at 18h

October 6 at 3h
November 6 at 1h
December 6 at 23h
January 6 at 21h
February 6 at 19h

WEST

EAST

SOUTH

ZENITH

AQR
PEG
Algenib
AND
Alph
PSC
M33
TRI
ARI
Mira
CET
FOR
ERI
PER
Algol
M34
TAU
Aldebaran
Capella
M38
M36
M37
M35
M1
AUR
Bellatrix
Rigel
ORI
M42
Saiph
Betelgeuse
M79
LEP
CMA
Sirius
M41
MON
Castor
Pollux
GEM
CMI
Procyon
CNC
M44
HYA

Southern Hemisphere Star Charts

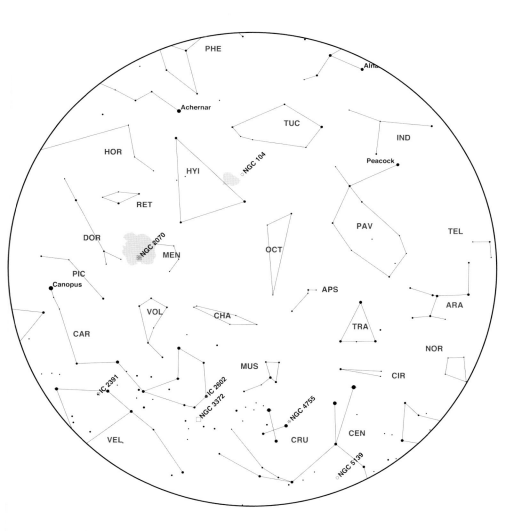

This chart shows stars lying at declinations between −45 and −90 degrees. These constellations are circumpolar for observers in Australia and New Zealand.

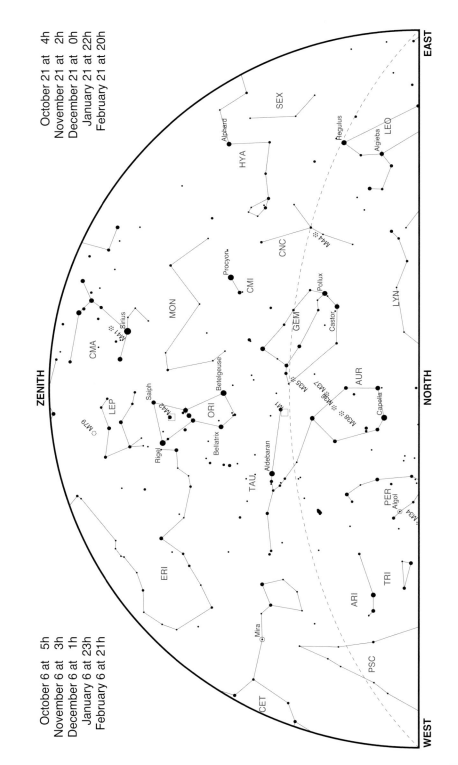

1N

October 6 at 5h
November 6 at 3h
December 6 at 1h
January 6 at 23h
February 6 at 21h

EAST

WEST

NORTH

ZENITH

SEX

HYA

Alphard

Regulus

LEO

Algieba

CNC

M44

Procyon

CMI

LYN

MON

Pollux

GEM

Castor

Sirius

M41

CMA

M35

AUR

Betelgeuse

M38

M36

M37

Capella

Saiph

LEP

M42

ORI

M79

Rigel

Bellatrix

M1

Aldebaran

TAU

PER

Algol

M34

ERI

ARI

TRI

Mira

PSC

CET

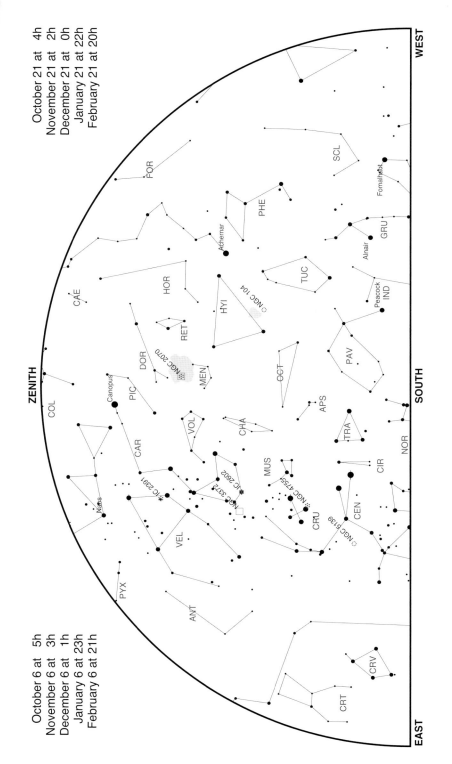

1S

WEST

EAST

ZENITH

SOUTH

October 21 at 4h
November 21 at 2h
December 21 at 0h
January 21 at 22h
February 21 at 20h

October 6 at 5h
November 6 at 3h
December 6 at 1h
January 6 at 23h
February 6 at 21h

FOR
SCL
PHE
Achernar
GRU
Alnair
CAE
HOR
TUC
HYI
NGC 104
Peacock
IND
RET
DOR
OCT
PAV
NGC 2070
MEN
COL
PIC
Canopus
CAR
VOL
CHA
APS
TRA
Sirius
IC 2391
NGC 3372
IC 2602
MUS
NOR
CIR
VEL
NGC 4755
CRU
NGC 5139
CEN
PYX
ANT
CRT
CRV
Fomalhaut

2N

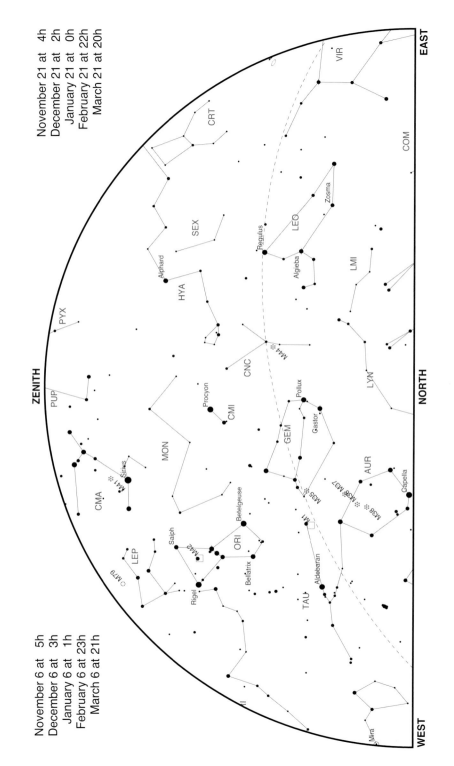

November 6 at 5h
December 6 at 3h
January 6 at 1h
February 6 at 23h
March 6 at 21h

EAST

VIR

CRT

COM

SEX

LEO

Zosma

Regulus

Algieba

LMI

HYA

Alphard

PYX

M44

CNC

LYN

ZENITH

PUP

Procyon

Pollux

CMI

GEM

Gastor

MON

AUR

Sirius

M41

CMA

M35

M38 M36 M37

Capella

M1

Betelgeuse

Saiph

M42

ORI

Aldebaran

LEP

Bellatrix

TAU

M79

Rigel

NORTH

Mira

WEST

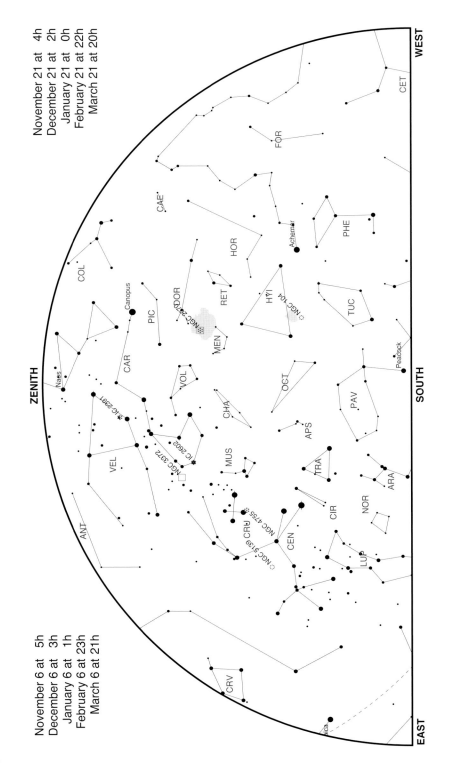

November 6 at 5h
December 6 at 3h
January 6 at 1h
February 6 at 23h
March 6 at 21h

November 21 at 4h
December 21 at 2h
January 21 at 0h
February 21 at 22h
March 21 at 20h

2S

WEST

ZENITH

SOUTH

EAST

CET

FOR

CAE

HOR

Achernar

PHE

COL

Canopus

PIC

DOR

NGC 2070

RET

HYI

NGC 104

TUC

CAR

MEN

VOL

OCT

Naos

IC 2391

CHA

PAV

Peacock

VEL

NGC 3372

IC 2602

MUS

APS

ANT

NGC 4755

CRU

CEN

TRA

ARA

NGC 5139

CIR

NOR

LUP

CRV

Spica

3N

January 21 at 2h
February 21 at 0h
March 21 at 22h
April 21 at 20h
May 21 at 18h

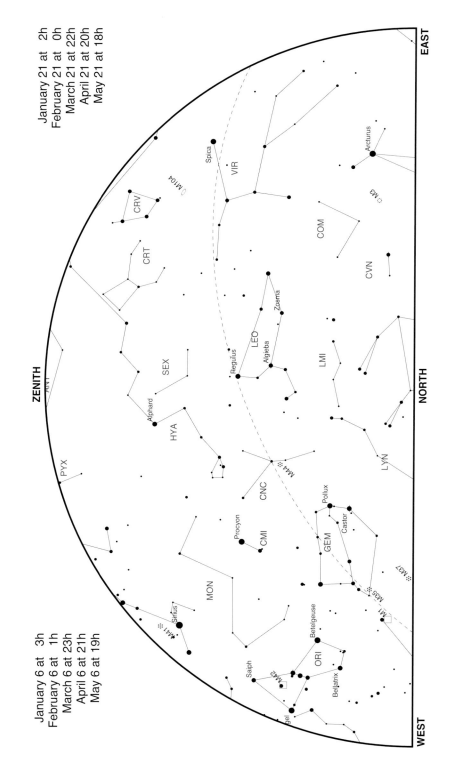

January 6 at 3h
February 6 at 1h
March 6 at 23h
April 6 at 21h
May 6 at 19h

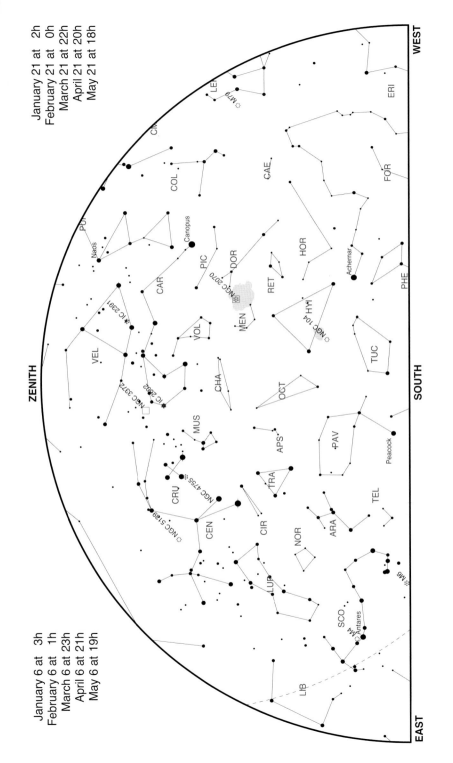

3S

WEST

ERI

FOR

PHE

SOUTH

ZENITH

EAST

LE

M79

CMa

COL

CAE

PUP

Naos

Canopus

PIC

DOR

HOR

CAR

NGC 2070

RET

MEN

NGC 104

Achernar

47 TUC

IC 2391

VOL

TUC

VEL

NGC 3372

CHA

IC 2602

OCT

MUS

APS

PAV

Peacock

CRU

NGC 4755

TRA

TEL

NGC 5139

CEN

CIR

NOR

ARA

M6

LUP

SCO

M4

Antares

LIB

4N

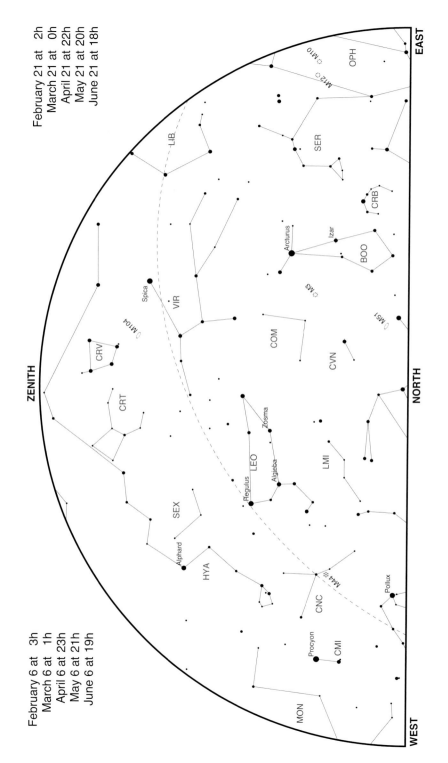

February 21 at 2h
March 21 at 0h
April 21 at 22h
May 21 at 20h
June 21 at 18h

February 6 at 3h
March 6 at 1h
April 6 at 23h
May 6 at 21h
June 6 at 19h

EAST

WEST

NORTH

ZENITH

OPH
M10
M12
SER
LIB
CRB
Izar
Arcturus
BOO
M3
M51
Spica
VIR
COM
CVN
M104
CRV
CRT
Zosma
LEO
Algieba
LMI
Regulus
SEX
Alphard
HYA
M44
CNC
Pollux
Procyon
CMI
MON

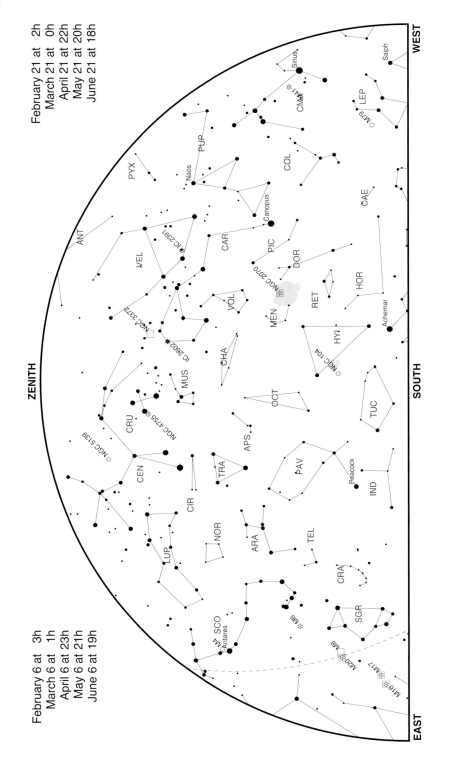

4S

February 21 at 2h
March 21 at 0h
April 21 at 22h
May 21 at 20h
June 21 at 18h

February 6 at 3h
March 6 at 1h
April 6 at 23h
May 6 at 21h
June 6 at 19h

WEST

EAST

SOUTH

ZENITH

5N

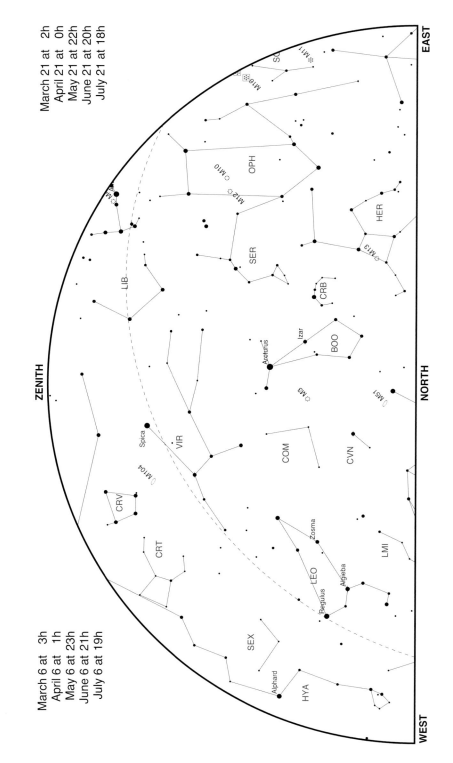

March 21 at 2h
April 21 at 0h
May 21 at 22h
June 21 at 20h
July 21 at 18h

March 6 at 3h
April 6 at 1h
May 6 at 23h
June 6 at 21h
July 6 at 19h

ZENITH

EAST

NORTH

WEST

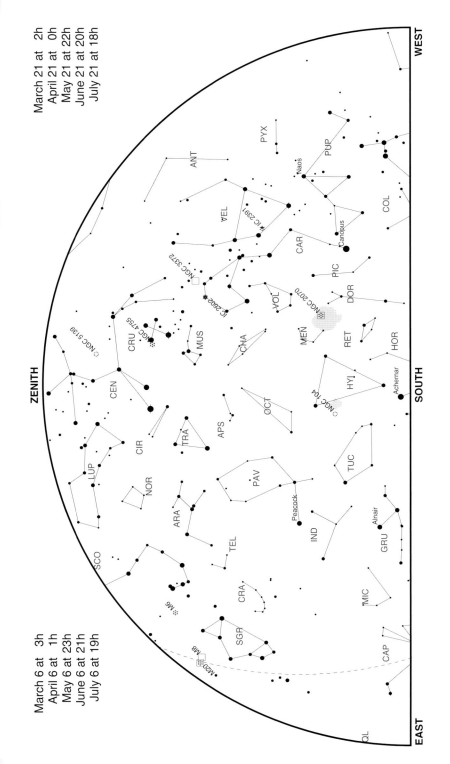

5S

March 6 at 3h
April 6 at 1h
May 6 at 23h
June 6 at 21h
July 6 at 19h

WEST

ZENITH

SOUTH

EAST

PYX
ANT
VEL
PUP
Naos
CAR
Canopus
COL
PIC
VOL
DOR
IC 2391
NGC 3372
IC 2602
NGC 2070
MEN
RET
NGC 5139
CRU
NGC 4755
MUS
CHA
HOR
Achernar
CEN
HYI
NGC 104
CIR
TRA
APS
OCT
LUP
NOR
PAV
TUC
ARA
IND
GRU
Alnair
SCO
TEL
Peacock
MIC
M6
CRA
SGR
CAP
M20 M8
QL

6N

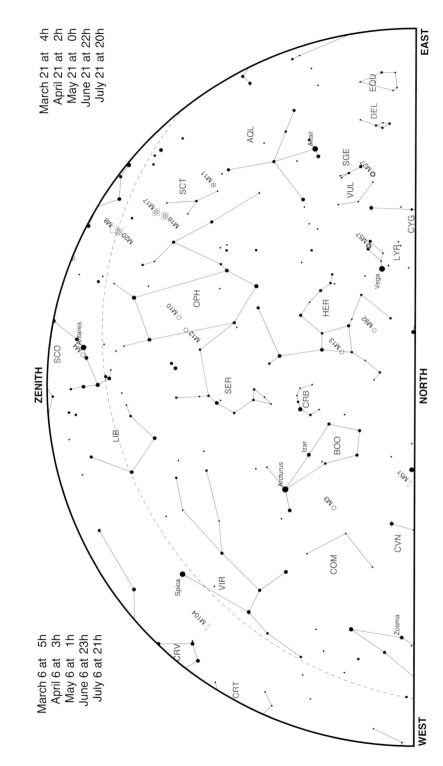

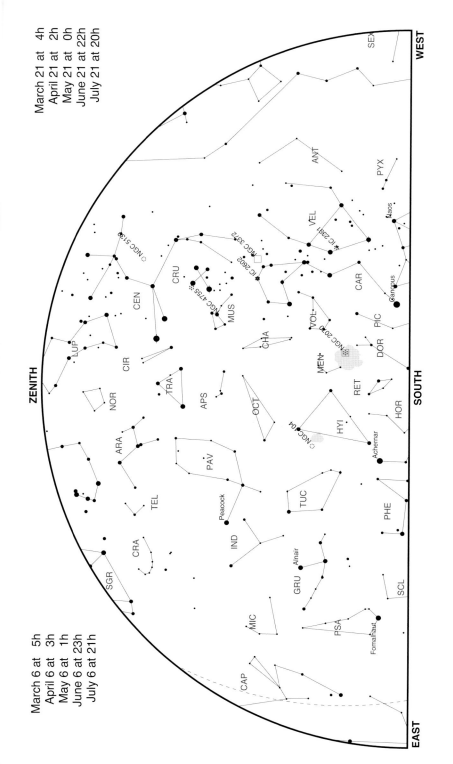

6S

WEST

March 21 at 4h
April 21 at 2h
May 21 at 0h
June 21 at 22h
July 21 at 20h

March 6 at 5h
April 6 at 3h
May 6 at 1h
June 6 at 23h
July 6 at 21h

ZENITH

SOUTH

EAST

SEX

PYX

ANT

VEL

CAR

PIC

DOR

RET

HOR

PHE

SCL

MEN

HYI

Achernar

TUC

GRU   Alnair

PSA

Fomalhaut

MIC

IND

PAV

Peacock

CAP

SGR

CRA

TEL

ARA

NOR

LUP

CIR

TRA

APS

OCT

CHA

MUS

CRU

CEN

VOL

Naos

Canopus

NGC 2391

IC 2602

NGC 3372

NGC 4755

IC 2602

NGC 5139

NGC 104

NGC 2070

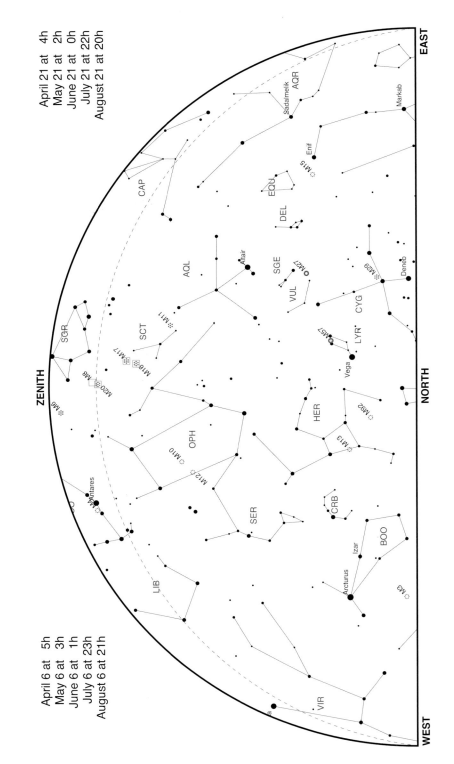

EAST

Sadalmelik
AQR
Markab
Enif
M15
EQU
DEL
CAP
SGE
Altair
AQL
VUL
M27
SCT
M11
CYG
Deneb
M29
SGR
M17
LYR
M57
Vega
ZENITH
M16
M8
M20
HER
M6
M92
OPH
M10
M13
M12
CRB
SER
BOO
Izar
Antares
Arcturus
M3
LIB
VIR
NORTH

WEST

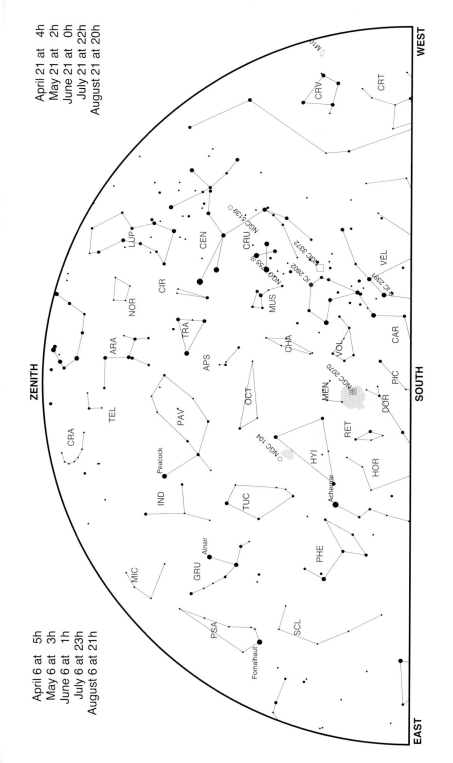

7S

WEST

EAST

ZENITH

SOUTH

April 21 at 4h
May 21 at 2h
June 21 at 0h
July 21 at 22h
August 21 at 20h

April 6 at 5h
May 6 at 3h
June 6 at 1h
July 6 at 23h
August 6 at 21h

CRV
CRT
VEL
IC 2391
CAR
PIC
DOR
RET
HOR
PHE
SCL
NGC 2070
MEN
HYI
Achernar
TUC
PSA
Fomalhaut
GRU
Alnair
MIC
IND
Peacock
PAV
TEL
CRA
ARA
NOR
CIR
LUP
CEN
CRU
NGC 5139
NGC 55
MUS
IC 2602
NGC 3372
CHA
VOL
OCT
APS
TRA
NGC 104

8N

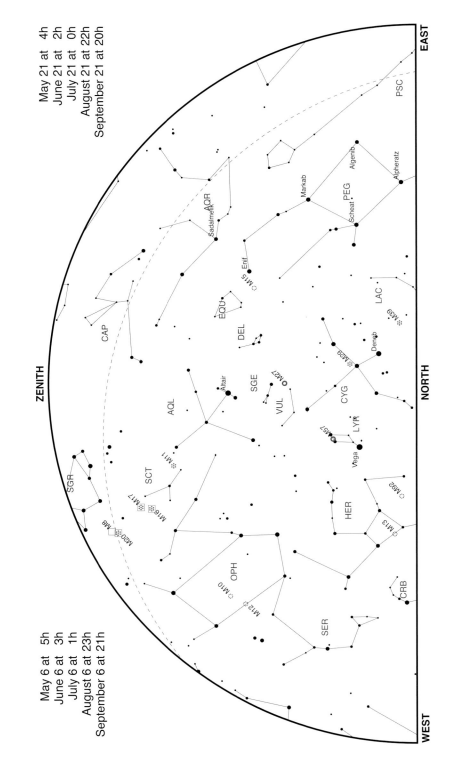

May 21 at 4h
June 21 at 2h
July 21 at 0h
August 21 at 22h
September 21 at 20h

May 6 at 5h
June 6 at 3h
July 6 at 1h
August 6 at 23h
September 6 at 21h

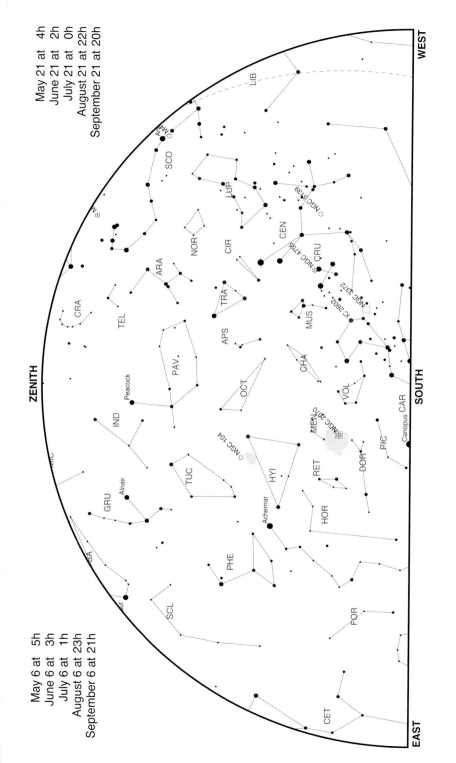

8S

WEST

ZENITH

SOUTH

EAST

May 21 at 4h
June 21 at 2h
July 21 at 0h
August 21 at 22h
September 21 at 20h

May 6 at 5h
June 6 at 3h
July 6 at 1h
August 6 at 23h
September 6 at 21h

LIB
SCO
LUP
NGC 5139
CEN
NOR
CIR
CRU
NGC 4755
ARA
TRA
IC 2602
NGC 4372
CRA
TEL
APS
MUS
NGC 3372
PAV.
CHA
VOL
IND
OCT
Peacock
MEN
NGC 2070
PIC
GRU
TUC
NGC 104
HYI
RET
DOR
Canopus CAR
Alnair
PHE
Achernar
HOR
SCL
FOR
CET

9N

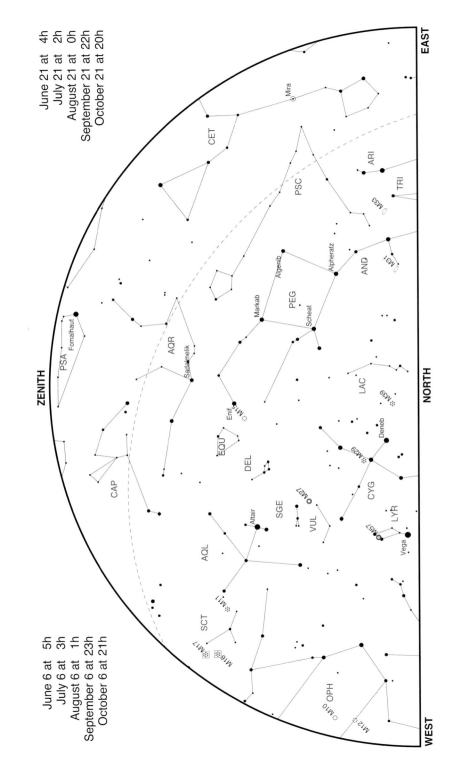

June 6 at 5h
July 6 at 3h
August 6 at 1h
September 6 at 23h
October 6 at 21h

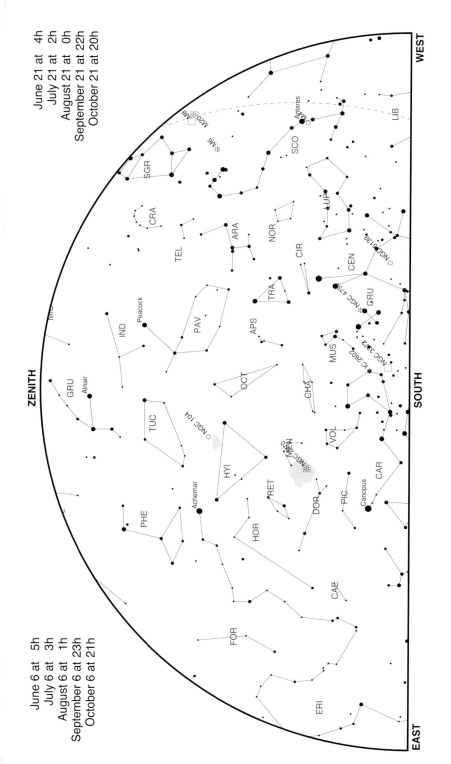

June 21 at 4h
July 21 at 2h
August 21 at 0h
September 21 at 22h
October 21 at 20h

June 6 at 5h
July 6 at 3h
August 6 at 1h
September 6 at 23h
October 6 at 21h

WEST

EAST

ZENITH

SOUTH

LIB

SCO

Antares

M7

M6

M20 M8

SGR

CRA

TEL

ARA

NOR

CIR

LUP

NGC 5139

CEN

TRA

NGC 4755

GRU

APS

MUS

IC 2602

NGC 3072

VOL

CHA

OCT

PAV

IND

Peacock

MIC

GRU

Alnair

TUC

NGC 104

HYI

Achernar

PHE

RET

HOR

NGC 2070

DOR

PIC

CAR

Canopus

CAE

FOR

ERI

10N

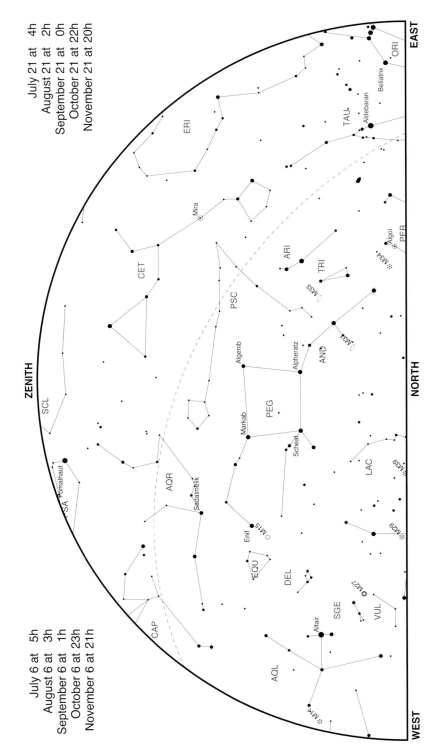

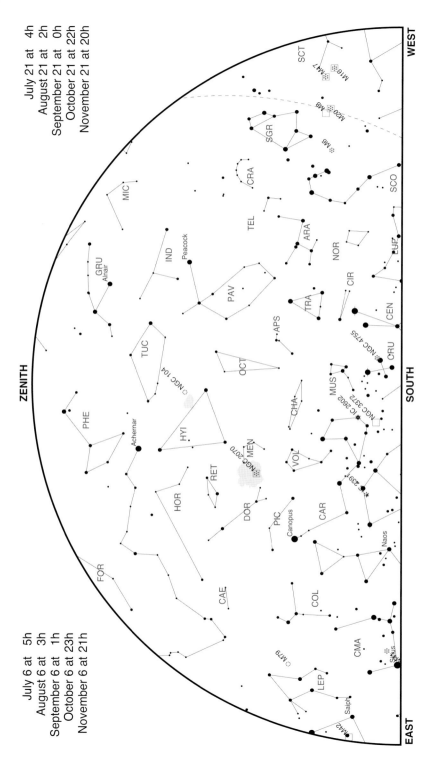

10S

July 21 at 4h
August 21 at 2h
September 21 at 0h
October 21 at 22h
November 21 at 20h

July 6 at 5h
August 6 at 3h
September 6 at 1h
October 6 at 23h
November 6 at 21h

ZENITH

WEST

SOUTH

EAST

SCT
M16
M17
M20
M8
M6
SGR
SCO
LUP
CRA
TEL
MIC
ARA
NOR
IND
Peacock
PAV
GRU
Alnair
APS
TRA
CIR
CEN
TUC
OCT
NGC 104
CRU
NGC 4755
MUS
IC 2602
NGC 3372
PHE
Achernar
HYI
CHA
MEN
NGC 2070
RET
VOL
2391
DOR
CAR
HOR
PIC
Canopus
Naos
FOR
CAE
COL
CMA
M79
Sirius
LEP
Saiph
M42

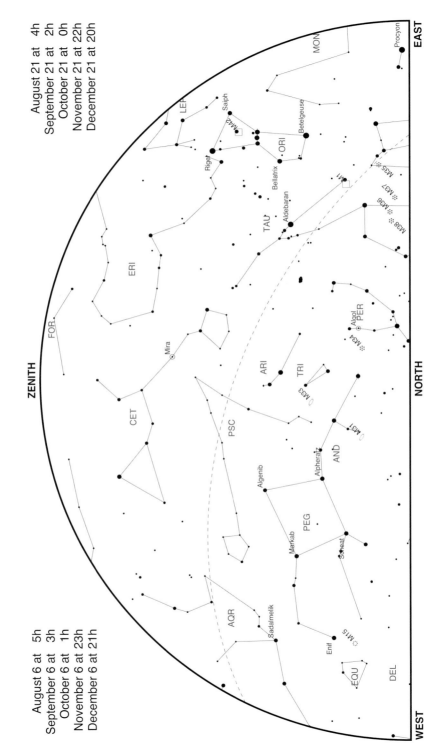

11N

August 6 at 5h
September 6 at 3h
October 6 at 1h
November 6 at 23h
December 6 at 21h

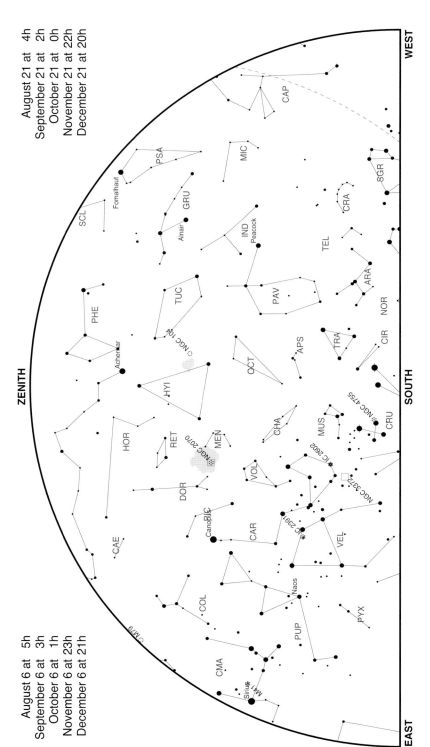

11S

August 21 at 4h
September 21 at 2h
October 21 at 0h
November 21 at 22h
December 21 at 20h

August 6 at 5h
September 6 at 3h
October 6 at 1h
November 6 at 23h
December 6 at 21h

WEST

ZENITH

SOUTH

EAST

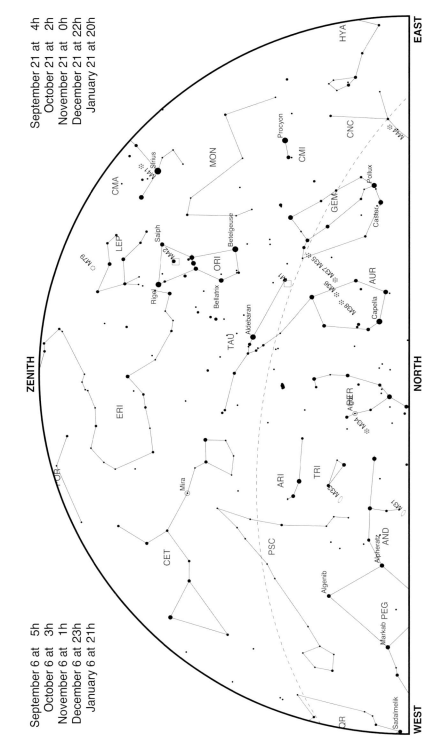

12N

September 6 at 5h
October 6 at 3h
November 6 at 1h
December 6 at 23h
January 6 at 21h

EAST

ZENITH

NORTH

WEST

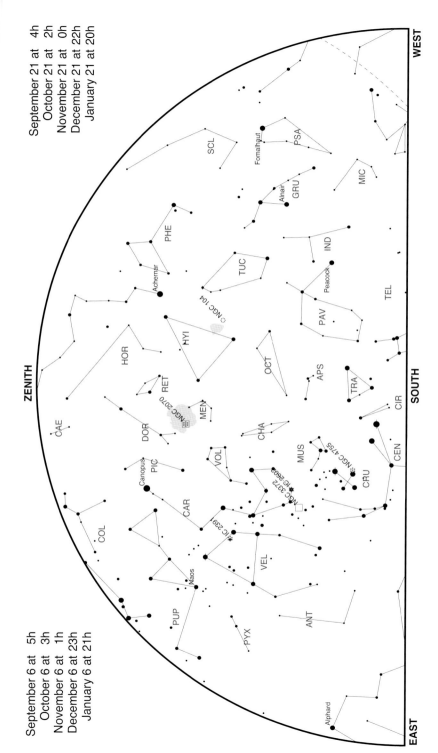

12S

September 21 at 4h
October 21 at 2h
November 21 at 0h
December 21 at 22h
January 21 at 20h

September 6 at 5h
October 6 at 3h
November 6 at 1h
December 6 at 23h
January 6 at 21h

WEST

ZENITH

SOUTH

EAST

SCL
PHE
Achernar
HOR
CAE
RET
DOR
PIC
Canopus
NGC 2070
MEN
CAR
COL
Naos
PUP
PYX
VEL
IC 2391
NGC 3372
VOL
CHA
MUS
CRU
IC 2602
NGC 4755
CEN
CIR
TRA
APS
PAV
OCT
HYI
NGC 104
TUC
IND
Peacock
TEL
MIC
GRU
Alnair
PSA
Fomalhaut
ANT
Alphard

The Planets in 2023

Lynne Marie Stockman

Mercury rapidly flits between the evening and morning skies several times each year. Evening apparitions start at superior conjunction and begin bright, ending with Mercury at sixth-magnitude. This year, the best evening appearance of Mercury for observers in northern temperate latitudes begins in mid-March and lasts until the end of April. July through early September offers the best evening viewing opportunities for astronomers in the southern hemisphere. Morning appearances of the tiny planet start faint and steadily brighten as Mercury leaves inferior conjunction. The May–June morning apparition favours southern latitudes whilst those in the northern hemisphere have their best dawn views between September and mid-October. In 2023, Mercury is occulted once by the Moon and encounters each of the planets, meeting Mars twice toward the end of the year. In July, Mercury is only 0.1° from first-magnitude star Regulus (α Leonis). *Apparition diagrams showing the position of Mercury above the eastern and western horizons can be found throughout the Monthly Sky Notes.*

Venus is the evening star for the first half of the year. It is a fair apparition for everyone, perhaps favouring observers in the tropics, with greatest elongation east taking place in June. With inferior conjunction occurring in mid-August, Venus ends the year in the morning sky and is best viewed from the northern hemisphere. The morning star reaches greatest elongation west in late October. Venus is at its brightest in July and September and faintest at the beginning of the year. It is twice occulted by the Moon this year and is found in conjunction with every other planet except Mars. It also passes near by Regulus twice, in July and October. In addition, Venus moves past two famous open star clusters, M45 (Pleiades) in April and M44 (Beehive or Praesepe) two months later. *Apparition diagrams showing the position of Venus above the western and eastern horizons can be found in the January and August Sky Notes respectively.*

The **Moon** embarks on a busy schedule of occulting both solar system objects and bright stellar bodies. Only Saturn escapes our satellite this year. Antares (α Scorpii), the brightest star in the constellation of the Scorpion, undergoes a lunar occultation on 25 August. It will continue to be blocked every month until August 2028. The Pleiades also falls victim to the Moon this year, with the first occultation taking place on 5 September. This series of lunar occultations of the open star cluster will continue until July 2029. Only four eclipses take place this year; details are in the article *Eclipses in 2023*.

Mars does not come to opposition this year. It begins 2023 at a bright magnitude −1.2 but dims to second magnitude as it progresses toward conjunction late in the year. It is an evening sky object for most of 2023 and is most easily seen from the northern hemisphere for much of that time. Mars starts in Taurus before moving through the zodiac, across the constellation of Ophiuchus and then on to Sagittarius at the end of December. It is occulted by the Moon five times, visits the open cluster Praesepe in June, skims past the two first-magnitude stars Regulus and Spica (α Virginis), and encounters Mercury both sides of conjunction in November. *A finder chart showing the position of Mars throughout the first eight months of 2023 follows this article.*

Jupiter spends the first part of the year in Pisces, with a short foray into Cetus, before moving into Aries in mid-May where it remains for the rest of 2023. It is occulted by the Moon four times, although one of these events is only visible during the day. Jupiter and Venus pair up in the west after sunset in early March but a similar meeting with Mercury near the end of that month may be too close to conjunction to observe. Jupiter opens the year in the evening sky but vanishes in the glow of sunset late in February before appearing in the east in April. It is a morning sky object until July and reaches opposition in early November. In January, Jupiter reaches perihelion for the first time since March 2011. *A finder chart showing the position of Jupiter throughout 2023 follows this article.*

Saturn is in Capricornus as 2023 begins but moves into Aquarius in mid-February where it remains for the rest of the year. An evening sky object low in the west at the beginning of January, it undergoes conjunction in February and moves into the morning sky. It returns to the west before midnight in May–June, depending on the observer's latitude, and reaches opposition in late August. The ringed planet remains in the evening sky for the rest of the year. It has close encounters with Venus and Mercury in January and March respectively, but these will be difficult to observe given Saturn's proximity to the Sun at the time. *A finder chart showing the position of Saturn throughout 2023 follows this article.*

Uranus resides in the constellation of Aries for the entirety of 2023. The series of lunar occultations which began last year ends in February. Uranus also has close encounters with Venus in March and Mercury in June. It begins as an evening sky object before undergoing conjunction with the Sun in early May. After this the green ice giant moves into the morning sky for the next two months, finally rising before midnight in late July. Opposition takes place in mid-November. *A finder chart showing the position of Uranus throughout 2023 follows this article.*

Neptune begins the year in the constellation of Aquarius, moving into Pisces in early March, before retrograde motion brings it back to Aquarius in late November. The most distant planet in the solar system then returns to Pisces in mid-December. Neptune is an evening sky object at the outset of 2023, visible in the west until its conjunction with the Sun in March. It then appears solely in the morning sky until around May–June when it begins to rise before midnight. Opposition occurs in September after which the faint planet is visible from sunset. Neptune is occulted by the Moon three times this year and has a very close encounter with Venus in mid-February. *A finder chart showing the position of Neptune throughout 2023 follows this article.*

Mars

January to August 2023

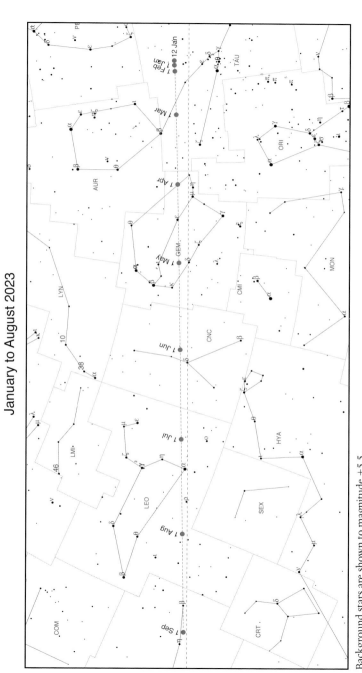

Background stars are shown to magnitude +5.5.

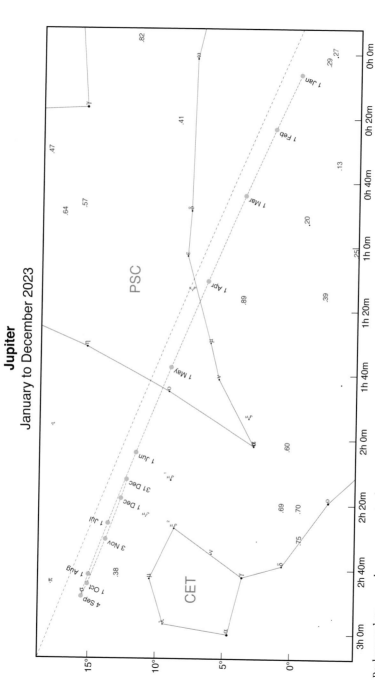

Jupiter
January to December 2023

PSC

CET

Background stars are shown to magnitude +5.5.

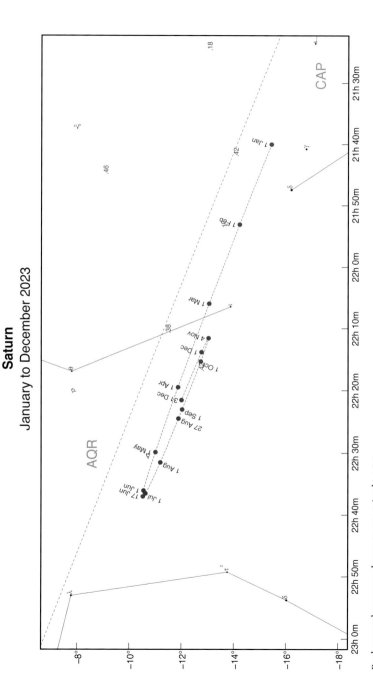

Saturn
January to December 2023

Background stars are shown to magnitude +5.5.

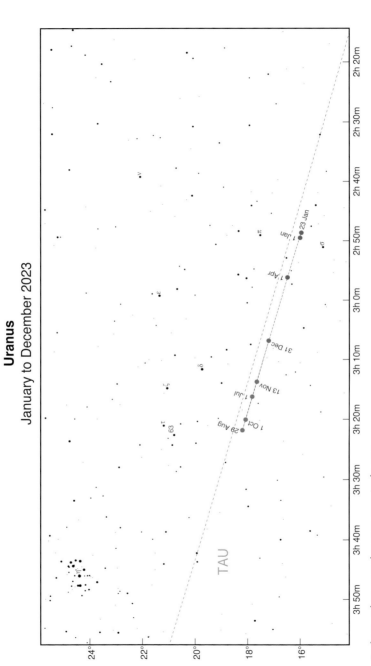

Uranus
January to December 2023

Background stars are shown to magnitude +8.0.

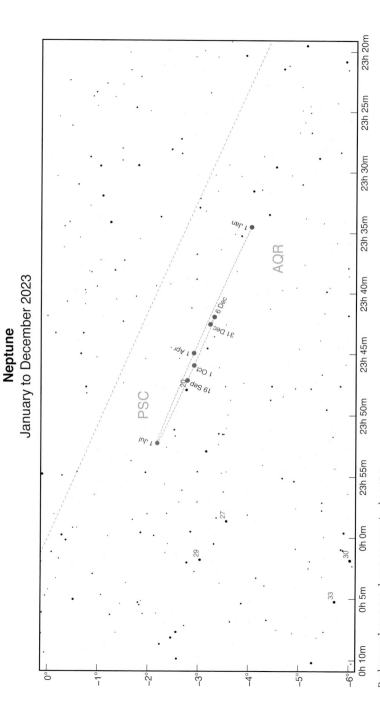

Neptune
January to December 2023

Background stars are shown to magnitude +10.0.

Phases of the Moon in 2023

New Moon	First Quarter	Full Moon	Last Quarter
		6 January	15 January
21 January	28 January	5 February	13 February
20 February	27 February	7 March	15 March
21 March	29 March	6 April	13 April
20 April	27 April	5 May	12 May
19 May	27 May	4 June	10 June
18 June	26 June	3 July	10 July
17 July	25 July	1 August	8 August
16 August	24 August	31 August	6 September
15 September	22 September	29 September	6 October
14 October	22 October	28 October	5 November
13 November	20 November	27 November	5 December
12 December	19 December	27 December	

Eclipses in 2023

There are a minimum of four eclipses in any one calendar year, comprising two solar eclipses and two lunar eclipses. Most years have only four, as is the case with 2023, although it is possible to have five, six or even seven eclipses during the course of a year. It is important to note that the times quoted for each event below refer to the start, maximum and ending of the eclipse on a global scale rather than with reference to specific locations. As far as lunar eclipses are concerned, these events are visible from all locations that happen to be on the night side of the Earth. Depending on the exact location of the observer, the entire eclipse sequence may be visible, although for some the Moon will be either rising or setting while the eclipse is going on.

The first eclipse of the year is the hybrid solar eclipse of 20 April. Hybrid solar eclipses are fairly uncommon and occur when the position of the Moon is such that the eclipse will appear as total to some observers and annular to others, the exact type of eclipse seen depending where on the path of totality the observer is located. Commencing in the southern Indian Ocean, this particular eclipse will move across parts of western Australia and southern Indonesia. The eclipse begins at 01:34 UT and ends at 06:59 UT with total eclipse taking place between 02:37 UT and 05:57 UT and maximum eclipse at 04:17 UT.

On 5 May there will be a penumbral lunar eclipse which will be visible throughout southern and eastern Europe, most of Asia, eastern Africa, Australia and New Zealand. The eclipse begins at 15:14 UT and ends at 19:31 UT with maximum eclipse occurring at 17:23 UT.

There will be an annular solar eclipse on 14 October, the path of which starts off the coast of southern Canada and moves across the south western United States and Central America, Columbia, Venezuela, Guyana, Suriname, French Guiana and Brazil. A partial eclipse will be visible throughout much of North and South America. The eclipse commences at 15:04 UT and ends at 20:55 UT. Full eclipse will last from 16:10 UT to 19:49 UT with maximum eclipse at 17:59 UT.

A partial lunar eclipse will take place on 28 October, the entirety or parts of which will be visible throughout North America, northern and eastern South America, Europe, Asia, Africa, most of Australia and parts of the Arctic and Antarctica. The eclipse commences at 18:02 UT and ends at 22:26 UT. The Moon begins to enter the Earth's umbra (full shadow) at 19:35 UT, with full eclipse lasting from 10:17 UT to 11:41 UT and maximum eclipse occurring at 20:14 UT. The Moon leaves the umbra at 20:52 UT.

Some Events in 2023

January	1	Saturn	Maximum ring opening (13.6°)
	1	Uranus	Lunar occultation (124° from the Sun)
	3	Mars	Lunar occultation (145° from the Sun)
	3 / 4	Earth	Quadrantid meteor shower (ZHR 120)
	4	Earth	Perihelion (0.9833 au)
	6	Full Moon	
	7	Mercury	Inferior conjunction (evening → morning)
	8	2 Pallas	Opposition in Canis Major
	15	Last Quarter Moon	
	20	Jupiter	Perihelion (4.9510 au)
	21	New Moon	Nearest perigee (356,570 km)
	22	Venus, Saturn	Planetary conjunction (0.3° apart, 22° from the Sun)
	28	First Quarter Moon	
	29	Uranus	Lunar occultation (96° from the Sun)
	30	Mercury	Greatest elongation west (25.0°)
	31	Mars	Lunar occultation (119° from the Sun)

February	4	Uranus	East quadrature
	5	Full Moon	Smallest Full Moon of the year (1,767 arc-seconds)
	13	Last Quarter Moon	
	15	Venus, Neptune	Planetary conjunction (0.01° apart, 28° from the Sun)
	16	Saturn	Conjunction
	20	New Moon	
	22	Jupiter	Lunar occultation (36° from the Sun)
	25	Uranus	Lunar occultation (69° from the Sun)
	27	First Quarter Moon	
	28	Mars	Lunar occultation (100° from the Sun)

March	2	Venus, Jupiter	Planetary conjunction (0.5° apart, 31° from the Sun)
	2	Mercury, Saturn	Planetary conjunction (0.9° apart, 12° from the Sun)
	7	Full Moon	
	15	Last Quarter Moon	
	15	Neptune	Conjunction
	16	Mercury, Neptune	Planetary conjunction (0.4° apart, 1° from the Sun)
	16	Mars	East quadrature
	17	Mercury	Superior conjunction (morning → evening)
	20	Earth	Equinox

March	21	1 Ceres	Opposition in Coma Berenices
	21	New Moon	
	22	Jupiter	Lunar occultation (15° from the Sun)
	24	Venus	Lunar occultation (35° from the Sun)
	29	First Quarter Moon	

April	6	Full Moon	
	11	Jupiter	Conjunction
	11	Mercury	Greatest elongation east (19.5°)
	13	Last Quarter Moon	
	19	Jupiter	Lunar occultation (6° from the Sun)
	20	New Moon	Hybrid solar eclipse
	22/23	Earth	Lyrid meteor shower (ZHR 18)
	27	First Quarter Moon	
	28	Moon	Nearest apogee (404,300 km)

May	1	Mercury	Inferior conjunction (evening → morning)
	5	Full Moon	Penumbral lunar eclipse
	6/7	Earth	Eta Aquariid meteor shower (ZHR 30)
	9	Uranus	Conjunction
	12	Last Quarter Moon	
	17	Jupiter	Lunar occultation (26° from the Sun)
	19	New Moon	
	27	First Quarter Moon	
	28	Saturn	West quadrature
	29	Mercury	Greatest elongation west (24.9°)
	30	Mars	Aphelion (1.6659 au)

June	2	Mars	0.1° north of M44 (Beehive Cluster or Praesepe)
	4	Full Moon	
	4	Venus	Greatest elongation east (45.4°)
	9	Earth	Tau Herculid meteor shower (ZHR varies)
	10	Last Quarter Moon	
	13	Saturn	Minimum ring opening (7.3°)
	13	Venus	0.5° north of M44 (Beehive Cluster or Praesepe)
	18	New Moon	
	19	Neptune	West quadrature
	21	Earth	Solstice
	26	First Quarter Moon	

July	1	Mercury	Superior conjunction (morning → evening)
	3	Full Moon	
	6	Earth	Aphelion (1.0167 au)
	7	15 Eunomia	Opposition in Sagittarius
	10	Last Quarter Moon	
	10	Mars	0.6° north of α Leonis (Regulus)
	14	Mercury	0.2° north of M44 (Beehive Cluster or Praesepe)
	17	New Moon	
	22	134340 Pluto	Opposition near the Capricornus/Sagittarius boundary
	25	First Quarter Moon	
	28	Mercury	0.1° south of α Leonis (Regulus)
	28/29	Earth	Delta Aquariid meteor shower (ZHR 20)

August	1	Full Moon	
	7	Jupiter	West quadrature
	8	Last Quarter Moon	
	10	Mercury	Greatest elongation east (27.4°)
	12/13	Earth	Perseid meteor shower (ZHR 80)
	13	Venus	Inferior conjunction (evening → morning)
	16	Uranus	West quadrature
	16	New Moon	Most distant apogee (406,635 km)
	24	First Quarter Moon	
	25	α Scorpii (Antares)	Lunar occultation (98° from the Sun)
	27	Saturn	Opposition in Aquarius
	31	Full Moon	Largest Full Moon of the year (2,006 arc-seconds)

September	1	Neptune	Lunar occultation (162° from the Sun)
	5	M45 (Pleiades)	Lunar occultation (104° from the Sun)
	6	Mercury	Inferior conjunction (evening → morning)
	6	Last Quarter Moon	
	15	New Moon	
	16	Mars	Lunar occultation (19° from the Sun)
	19	Neptune	Opposition in Pisces
	21	α Scorpii (Antares)	Lunar occultation (72° from the Sun)
	22	Mercury	Greatest elongation west (17.9°)
	22	First Quarter Moon	
	23	Earth	Equinox
	28	Neptune	Lunar occultation (170° from the Sun)
	29	Full Moon	'Harvest Moon'

October	3	M45 (Pleiades)	Lunar occultation (130° from the Sun)
	6	Last Quarter Moon	
	8/9	Earth	Draconid meteor shower (ZHR 10)
	10	Earth	Southern Taurid meteor shower (ZHR 5)
	14	Mercury	Lunar occultation (4° from the Sun)
	14	New Moon	Annular solar eclipse
	15	Mars	Lunar occultation (11° from the Sun)
	18	α Scorpii (Antares)	Lunar occultation (45° from the Sun)
	20	Mercury	Superior conjunction (morning → evening)
	21/22	Earth	Orionid meteor shower (ZHR 20)
	22	First Quarter Moon	
	23	Venus	Greatest elongation west (46.4°)
	28	Full Moon	Partial lunar eclipse
	29	Mercury, Mars	Planetary conjunction (0.3° apart, 6° from Sun)
	30	M45 (Pleiades)	Lunar occultation (157° from the Sun)

November	3	Jupiter	Opposition in Aries
	5	Last Quarter Moon	
	9	Venus	Lunar occultation (46° from the Sun)
	12	Earth	Northern Taurid meteor shower (ZHR 5)
	13	New Moon	
	13	Uranus	Opposition in Aries
	14	α Scorpii (Antares)	Lunar occultation (18° from the Sun)
	17/18	Earth	Leonid meteor shower (ZHR varies)
	18	Mars	Conjunction
	20	First Quarter Moon	
	21	Moon	Most distant perigee (369,824 km)
	23	Saturn	East quadrature
	27	M45 (Pleiades)	Lunar occultation (174° from the Sun)
	27	Full Moon	

December	4	Mercury	Greatest elongation east (21.3°)
	5	Last Quarter Moon	
	12	α Scorpii (Antares)	Lunar occultation (10° from the Sun)
	12	New Moon	
	13/14	Earth	Geminid meteor shower (ZHR 75+)
	17	Neptune	East Quadrature
	18	37 Fides	Opposition in Auriga
	19	Neptune	Lunar occultation (87° from the Sun)
	19	First Quarter Moon	
	21	4 Vesta	Opposition in Orion
	22	Earth	Solstice
	22	Mercury	Inferior conjunction (evening → morning)
	22/23	Earth	Ursid meteor shower (ZHR 10)
	24	M45 (Pleiades)	Lunar occultation (147° from the Sun)
	27	Full Moon	

Note that the dates quoted in the calendar are based on UT and that events listed may occur one day before or after the given date depending on the observer's time zone.

The lunar occultations of M45 (Pleiades) refer to the occultation of the brightest member of the cluster, Alcyone (η Tauri).

For more on each of the eclipses occurring during the year, please refer to the information given in *Eclipses in 2023*.

The entries relating to meteor showers state the *estimated* date of peak shower activity (maximum). The figure quoted in parentheses in column 4 alongside each meteor shower entry is the expected Zenith Hourly Rate (ZHR) for that particular shower at maximum. For a more detailed explanation of ZHR, and for further details of the individual meteor showers listed here, please refer to *Meteor Showers in 2023*.

Minor planet oppositions are measured with respect to ecliptic longitude; the dates of opposition in right ascension and greatest elongation from the Sun may differ by several days, depending on the orbital characteristics of the object. Please see the *Monthly Sky Notes* and *Minor Planets in 2023* for more details.

Monthly Sky Notes and Articles

Evening Apparition of Venus
October 2022 to August 2023

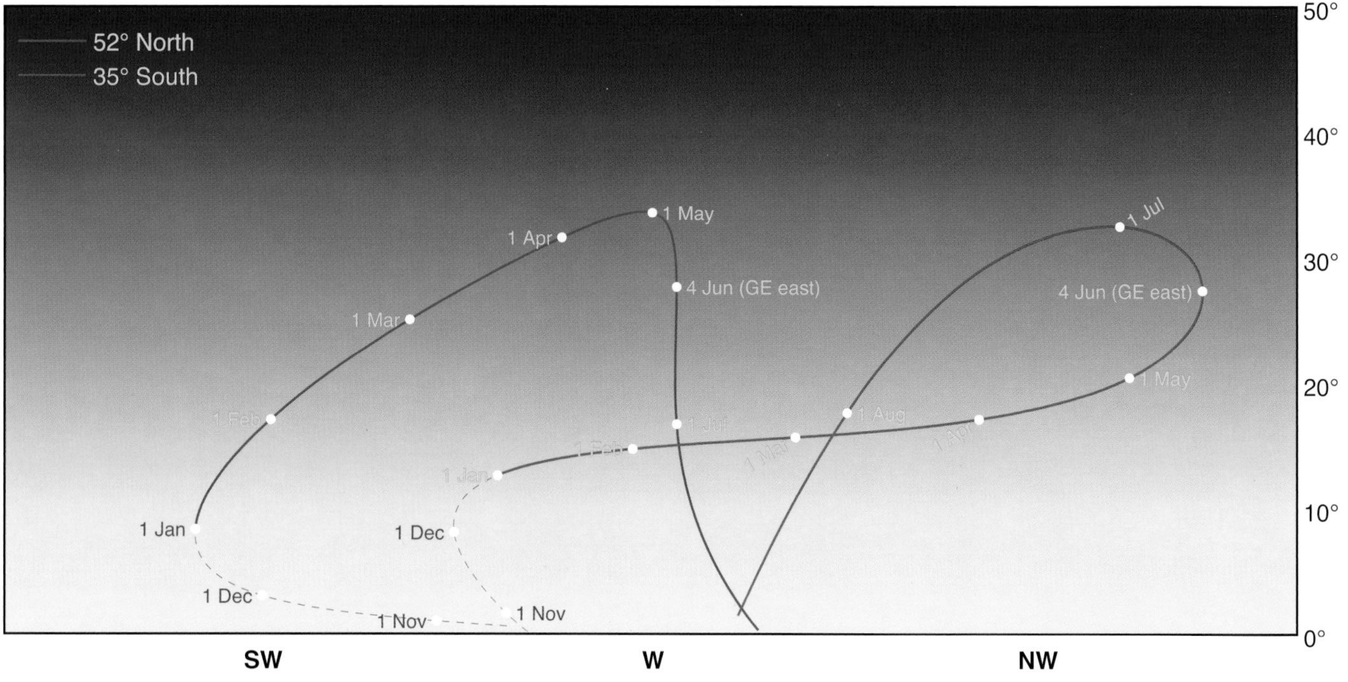

52° North
35° South

50°
40°
30°
20°
10°
0°

NW
W
SW

1 Jul
4 Jun (GE east)
1 May
1 Aug
1 May
4 Jun (GE east)
1 Apr
1 Mar
1 Feb
1 Jan
1 Dec
1 Dec
1 Nov
1 Nov
1 Jan
1 Dec
1 Jan

January

Full Moon: 6 January
Last Quarter: 15 January
New Moon: 21 January
First Quarter: 28 January

MERCURY reaches the first of five perihelia in 2023 on the second day of January. This tiny planet is briefly an evening sky object but is very low in the west and soon disappears, undergoing inferior conjunction on 7 January and moving to the morning sky. Early risers in the southern hemisphere and the tropics see Mercury vault high above the eastern horizon whilst northern temperate observers are not so favoured. After entering the year in retrograde, Mercury returns to direct or prograde motion on 18 January, and attains greatest elongation west (25.0°) on the penultimate day of the month. It is best seen late in January when it is at its highest and brightest.

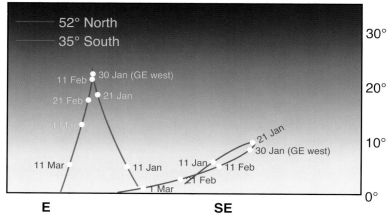

Morning Apparition of Mercury
7 January to 17 March

VENUS was at superior conjunction last October and is currently visible in the evening sky, gaining in altitude every night. It is over 1.5 au away from Earth but is slowly catching up. Venus appears 0.3° north of a much dimmer Saturn on 22 January. When the two bodies are viewed through a telescope, Venus appears as a waning gibbous orb a little under 11 arc-seconds across and shining at magnitude −3.9. The disk of Saturn, by comparison, is nearly 16 arc-seconds in diameter but the ringed planet is only magnitude +0.8.

EARTH reaches perihelion on 4 January when it is at a distance of 0.9833 au from the Sun. The Quadrantid meteor shower also peaks about this time but is washed out by the nearly Full Moon. See *Meteor Showers in 2023* for more details. The Moon has a busy month, occulting both Uranus and Mars twice. The nearest perigee of the year coincides with New Moon on 21 January which brings about a proxigean spring tide, a rare and extreme form of a spring tide.

MARS executes a pirouette just north of the 'head' of the Bull, Taurus, after starting out the year in retrograde and then resuming direct motion mid-month. It is twice occulted by the Moon this month. The first occurrence takes place on 3 January beginning around 18:00 UT which is visible in the southern half of Africa, plus Madagascar. The western hemisphere gets its chance on the last day of the month when the waxing gibbous Moon obscures Mars starting at approximately 2:00 UT; this event is visible from the southern United States, Mexico, Central America, the Caribbean islands and the north western corner of South America. Mars is visible in evening skies and is best viewed from northern latitudes where it remains above the horizon most of the night. *A finder chart showing the position of Mars throughout the first eight months of 2023 follows the article The Planets in 2023.*

2 PALLAS is at opposition in ecliptic longitude on 8 January. Opposition in right ascension occurs five days earlier on 5 January and greatest elongation from the Sun takes place on 16 January when the asteroid is 126° from the Sun. 2 Pallas has a highly eccentric (0.23) and inclined (35°) orbit which accounts for the wide variation in these dates. This eighth-magnitude asteroid is located in Canis Major; see *Minor Planets in 2023* for more information.

JUPITER spends the month in the sprawling constellation of Pisces. It is an evening sky object and sets before midnight, but is well-positioned for viewing in both hemispheres. On 20 January, Jupiter comes to perihelion when it is 4.95 au from the Sun; it will be another 12 years before this happens again. The waxing crescent

Moon comes to call on 26 January, passing less than 2° south of the magnitude −2.3 gas giant.

SATURN is in the constellation of Capricornus as the year commences, a little over a degree north of the fourth-magnitude star Nashira (γ Capricorni). The rings are open to 13.6°, their maximum for the year. Around mid-month, Saturn passes 1.4° north of Deneb Algedi (δ Capricorni) and on 22 January, is found 0.3° south of the evening star. Saturn is an evening sky object, low in the west and vanishing in the twilight by the end of the month.

URANUS begins the year with a lunar occultation on the first day of the month. Visible beginning around 20:00 UT, observers from the north eastern United States, northern Canada, the Arctic regions including Greenland and Iceland, most of the British Isles, Scandinavia and north western Russia will have the opportunity to see the waxing gibbous Moon pass in front of the sixth-magnitude planet. A second lunar occultation takes place on 29 January; it begins around 2:00 UT and again is visible from the Arctic, this time including Alaska and northern Siberia along with Greenland. Having entered the year in retrograde, Uranus returns to direct motion on 23 January. It is found in the constellation of Aries and is visible in the evening hours. Northern hemisphere astronomers have the best views, where the ecliptic is high overhead.

NEPTUNE is visible in the west after sunset. The eighth-magnitude object inhabits the constellation of Aquarius and is best viewed in the middle of the month when the Moon is absent from evening skies. The waxing crescent Moon passes less than 3° south of the planet on 25 January.

Graduate Students and the Nobel Prize

David M. Harland

The ultimate accolade for scientific achievement is widely regarded as the Nobel Prize. But how senior must someone be to be worthy? In particular, should it be shared with research students who participated in momentous discoveries? There are two notable cases.

-oOo-

In seeking planets by how they perturb their stars, it is necessary to make exceedingly fine radial velocity measurements. In 1993 an echelle spectrograph with a diffraction grating optimised for high dispersion of spectral features was installed on the 1.93-metre telescope at the Haute Provence Observatory in France.

At a conference in Florence, Italy, in October 1995 Michel Mayor and Didier Queloz of the University of Geneva announced that there was a planet orbiting the star 51 Pegasi. Queloz was working for his doctorate under the supervision of Mayor, and his task was to analyse the data. The plan was to monitor 142 G-type and K-type stars for indications of gravitational perturbations. They started taking data in late 1994 and by March 1995 it was evident there was a periodicity in

Didier Queloz (left) and Michel Mayor with the 3.6 metre telescope at La Silla in the background. (ESO/L. Weinstein/Ciel et Espace Photos)

The presence of a planet produces a cyclical variation in the radial velocity of the star 51 Pegasi. The data has been folded into a 'phase' diagram with a period of 4.2 days. This is Figure 4 of 'A Jupiter-Mass Companion to a Solar-Type Star', M. Mayor and D. Queloz, *Nature*, **378**, 355–359, 1995.

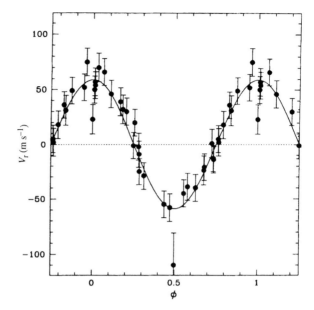

51 Pegasi. What was suspicious was the cycle of only 4.2 days. Rather than publish, they waited for the star to become visible again in July in order to verify that the signal was still in phase, which it was.[1]

The irony was that Mayor and Queloz were not seeking planets at all. They were interested in objects too large to be planets and too small to be stars, the so-called 'brown dwarfs' with masses between 15 and 80 times that of Jupiter. Since they would be difficult to detect in isolation, they decided to look for them orbiting normal stars. It was simple good luck that they found a planet.

With a G5 spectral type, 51 Pegasi is slightly cooler than the Sun but it is more luminous because its radius is larger. The mass of the planet is about half that of Jupiter. Because its orbit is a mere 0.05 au in radius the planet has been described as a 'hot Jupiter'.

In 2019 Mayor and Queloz were awarded the Nobel Prize for Physics for this discovery.[2] As noted, Queloz was a student at the time the work was carried out.

---

1. 'A Jupiter-Mass Companion to a Solar-Type Star', M. Mayor and D. Queloz, *Nature*, **378**, 355–359, 1995.
2. In fact, Mayor and Queloz shared the Nobel Prize with P. J. E. Peebles, the Princeton theorist who has produced many insights into the nature of the universe unrelated to extrasolar planets.

In contrast, in 1965 Jocelyn Bell started her doctoral research under Antony Hewish at Cambridge University by stringing 2,048 full-wave dipoles in 16 rows of 128 elements to form a novel radio telescope at the Mullard Radio Astronomy Observatory. The plan was to measure the angular diameters of celestial radio sources by the way in which they scintillated. Once the telescope was operating, it was her task to analyse the data. She started making observations in the summer of 1967 and by the end of the year had discovered several rapidly pulsating radio sources that would become known as pulsars.

In 1974 the Nobel Prize for Physics was awarded, in part, to Hewish "for his decisive role in the discovery of pulsars".[3]

In his acceptance speech, Hewish explained the key contribution to his research made by Bell. In the journal paper reporting the discovery, her name was second, behind his own in the list of five authors, so she had received the appropriate scientific credit.[4]

When Fred Hoyle was on a lecture tour of Canada the following year he was asked by journalist whether he thought Bell should have been included in the Nobel. The rules permit a maximum of three recipients, so it was not as if Bell had been nudged out, she had simply not been assigned the third slot.

Hoyle's response was impromptu,[5] "Yes, Jocelyn Bell was the actual discoverer, not Hewish, who was her supervisor, so she should have been included." This was reported in the *Montreal Gazette* on 21 March in an article by André Potworowski. Although Hoyle had simply said Bell should have been included because she was, strictly speaking, the discoverer, the spin of the story was that Hewish had *stolen* the credit from Bell.

When asked by the *Daily Mail* in London, Hewish was appalled,[6] "What Hoyle has said is untrue, absolutely untrue. It makes me so angry. It is ridiculous to suggest that the results were stolen."

In response to the furore, Bell noted,[7] "It was Professor Hewish who did the preparatory work. I was the individual who analysed the data, doing the spade work."

---

3. Hewish shared the prize with his boss, Martin Ryle, for other contributions to radio astronomy.
4. 'Observations of a Rapidly Pulsating Radio Source', A. Hewish, S. J. Bell, J. D. H. Pilkington, P. F. Scott and R. A. Collins, *Nature*, **217**, 709–713, 1968.
5. Fred Hoyle: *A Life In Science*, Simon Mitton, Aurum Press, p. 302, 2005.
6. *ibid*, p. 302
7. *ibid*, p. 302

When Hoyle, now in America, heard of the reactions to his off-the-cuff remarks he was astounded. His final thought on the affair, written to *The Times* on 1 April, was the Bell hadn't been included in the award just because "the [Nobel] Committee did not bother itself to understand what happened."[8] In his view, Bell had not simply undertaken routine work, she had shown initiative. She had made a real contribution as a co-investigator and hence deserved recognition. *That* was why he had said she should have been included in the Nobel Prize.

Jocelyn Bell in 1967. (Roger W. Haworth / Wikipedia Creative Commons)

As Hewish said later,[9] Bell was "a diligent graduate student" but she had played no part in planning the project or designing the telescope, and once the discovery had been made she "did not have the expertise to initiate further observations, such as accurate timing, distance estimates, or Doppler measurements."

On being invited by Thomas Gold to deliver an after dinner speech at the Eighth Texas Symposium on Relativistic Astrophysics in 1977, Bell said,[10] "It is the supervisor who has the final responsibility for the success or failure of the project." She added, "I believe it would demean the Nobel Prize if they were to award to research students, except in very exceptional cases, and I do not believe this is one of them."

-oOo-

Having ignored Bell and recognised Queloz, it is good to see that in recent years the Royal Swedish Academy of Sciences has developed a more inclusive attitude.

8. *ibid*, p. 303
9. *Clocks in the Sky: The Story of Pulsars*, Geoff McNamara, Springer, p. 51, 2008.
10. *Petit Four* by S. Jocelyn Bell Burnell, an after dinner speech at the Eighth Texas Symposium on Relativistic Astrophysics, published in the *Annals of the New York Academy of Sciences*, **302**, 685–689, 1977.

February

Full Moon: 5 February
Last Quarter: 13 February
New Moon: 20 February
First Quarter: 27 February

MERCURY is losing altitude above the eastern horizon but remains visible throughout the month, with observers in the southern hemisphere getting the best morning views. The tiny planet continues to brighten, ending the month at magnitude −0.6. It reaches its descending node on 5 February and moves through aphelion ten days later.

VENUS is the evening star, slowly brightening from magnitude −3.9 to −4.0 over the course of the month. It is gaining altitude above the western horizon from night to night, more quickly when viewed from the northern hemisphere than from the south. A telescope will be necessary to view both Venus (approximately 90% illuminated and just under 12 arc-seconds across) and Neptune (only 2.4 arc-seconds in apparent diameter) on 15 February when the two objects are just 0.01° apart in evening twilight. The waxing crescent Moon is 2.1° south of the bright planet on 22 February.

EARTH witnesses three lunar occultations this month, of Jupiter, Uranus and Mars. The Full Moon of 5 February has the smallest apparent size of the year.

MARS shines at zero magnitude in Taurus. It appears as a red gibbous disk in a telescope and shrinks in apparent size (10.7 arc-seconds to 8.2 arc-seconds) as it gets farther from Earth. On the last day of the month, the waxing gibbous Moon occults Mars beginning around 4:00 UT, an event which is visible from northern Scandinavia, Svalbard, Iceland and the north eastern coast of Greenland. Mars is an evening sky object, best seen in the cold dark winter skies of the northern hemisphere.

JUPITER makes a two-week foray into the non-zodiacal constellation of Cetus this month, spending the rest of February in Pisces. Shining at magnitude −2.2, the giant planet is visible in the west after sunset but is gone by late evening.

On 22 February, the young crescent Moon occults Jupiter in an event beginning at 22:00 UT and visible from the Falkland Islands, Argentina, Uruguay, southern Paraguay and southern Brazil.

SATURN is largely lost to view this month. It departs Capricornus and enters Aquarius three days before conjunction on 16 February. The ringed planet is found just $0.8°$ north of the fourth-magnitude star Iota (ι) Aquarii on the penultimate day of February but both objects are deep in morning twilight at the time.

URANUS reaches east quadrature on 4 February. Found in Aries, the faint planet is an evening sky object, setting around midnight for observers in northern temperate regions but rather earlier for those enjoying southern summer skies. Uranus began a series of monthly lunar occultations in February 2022; this comes to an end on 25 February when the waxing crescent Moon blots out Uranus as seen from parts of Baffin Island in the far north. *A finder chart showing the position of Uranus throughout 2023 follows the article The Planets in 2023.*

NEPTUNE is low in the west as night falls and is lost in twilit skies before the end of the month. Located in Aquarius, the eighth-magnitude planet is found only $0.01°$ south of the evening star on 15 February but a telescope will be necessary to capture both objects in skies which may not be completely dark. The very young crescent Moon passes $2.5°$ south of the planet on 21 February.

Giant Leaps for Small Change

Carolyn Kennett

Did you know that there is a USA dime on Mars? and right at this moment two USA quarter's flying at high velocity in the outer Solar System, a journey that has already taken them past Pluto? These are two examples of minted coins that have become part of space missions.

Size and weight defines what can be carried on a spacecraft and the incorporation of small change within space missions is as duel faceted as the coins themselves. There is usually a practical reason for inclusion as well as a commemorative.

Florida State Quarter, one of a pair of USA quarters that are flying on the New Horizons spacecraft. (NASA / Bill Rogers, JHU / APL)

When the New Horizons space craft launched in 2006, onboard was not one but two USA state quarters – the first from Florida and the second from Maryland. The coins were chosen to honour the states where the craft was built and where it was launched from. The story goes that the idea for their inclusion came during a long drive undertaken by Dr. Alan Stern the New Horizons Principal Investigator and Jim Kennedy the Space Kennedy Director. On their way to see Governor Bush they discussed how they could make him more interested in the mission and came to the conclusion it would be great to include a state coin, a Florida quarter would be perfect. This is where the tale takes an interesting twist, on route to the meeting the pair decided to find one of the coins and pulled into a Burger King restaurant. "We tried to find a state quarter in their cash registers. We had their entire staff looking," Stern said "It was a pretty surreal scene. The entire time I was thinking, 'Here are these 18- and 19-year-old, minimum wage folks rifling for a quarter that's going to fly to the Kuiper belt." Unfortunately, their search was not successful with Stern and Kennedy continuing on their way empty handed. They didn't mention their idea until the end of the briefing and Stern recalls that, "I mentioned to (Governor Bush) that we really wanted to fly a Florida state quarter but couldn't come up with one and it was due to our poor planning because we had only thought of it today, and he said, 'Well, I've got plenty!' and with that ran out of the room and when he came back, he had a roll of the quarters. He said, 'Fly these!'"

The Florida coin would not be one of Bush's but a freshly designed "Gateway to Discovery" quarter. It was by mounted onto the craft at Kennedy Space Center's Payload Hazardous Servicing Facility and had a purpose of being a 'Spin Balance' weight, along with its counterpart the Maryland coin. It would add an essential few kilograms in the correct part of the spacecraft making it stable. These coins have now passed Pluto and in January 2019 New Horizons flew past the furthest object ever explored called 486958 Arrokoth located 42.7 au from the Sun, in the Kuiper Belt.

When Curiosity launched to Mars on 26 November 2011, onboard was a Lincoln penny. Originally due for launch in 2009 the penny would commemorate the bicentennial of Lincoln's birth and centennial of the Lincoln Cent. The coin had been included as part of the calibration target of the Mars Hand Lens Imager (MAHLI) instrument. This camera is located at the end of the rover's robotic arm and the coin acts as an object that people can use to easily perceive size, it is also a nod to geologists who had traditionally used coins in photographs to act as a scale reference.

Some coins have much shorter journeys into space. In May 2019 astronaut Christina Koch recorded a message from the International Space Station holding two 50th Anniversary lunar landing commemorative coins. The coins were returned to Earth in June and with one going on display in Smithsonian National Air and Space Museum and the other being returned to the Mint for safe keeping.

When the Hubble Space Telescope required repairing two astronauts from Ohio took four Ohio commemorative quarter-dollar coin's with them on the mission. Carried onboard the Space Shuttle Columbia, the freshly

Lincoln Coin as part of the calibration target for the Mars Hand Lens Imager (MAHLI) aboard NASA's Mars Rover Curiosity. (NASA/JPL-Caltech/Malin Space Science Systems)

The David Bowie one-ounce silver proof coin in space. (Royal Mint)

minted coins paid tribute to the men and women who advanced aviation and space exploration further than ever before, Ohio being the birthplace of many pioneers in this field, including Neil Armstrong.

The final example is a tribute to the English singer-songwriter David Bowie. The first ever UK coin launched into space was designed by the Royal Mint in honour of the life and career of the musician Bowie, who released multiple hit records under his alter ego, Ziggy Stardust including *Life on Mars?*, *Space Oddity* and *Starman*. A one-ounce silver proof coin journeyed for 45 minutes reaching 35,656 metres before descending back to Earth. If the coin could speak, we might ask if it sang the lines of Bowie's song *Space Oddity* on its journey. One would like to think so. After all, we do live on a beautiful blue planet.

You can watch the David Bowie's commemorative coin journey here **www.youtube.com/watch?v=u7u2nOVI5vQ**

A message from astronaut Christina Koch at the International Space station **www.youtube.com/watch?v=DUILx9eY0Ng**

March

Full Moon: 7 March
Last Quarter: 15 March
New Moon: 21 March
First Quarter: 29 March

MERCURY has three close encounters with superior planets this month. On the second day of March, Mercury passes 0.5° south of Saturn with the inferior planet much the brighter object. It has an even closer approach to Neptune on 15 March when the two bodies are 0.4° apart but with superior conjunction only two days later, this event is not visible. After moving into the evening sky, the smallest planet in the solar system is found 1.8° north of the very young crescent Moon on 22 March. A close encounter with Jupiter occurs low in the west on 28 March, the day after Mercury passes through its ascending node and the day before it reaches perihelion. This is the beginning of the best evening apparition of the year for astronomers in northern temperate latitudes.

VENUS continues to appear a little higher above the western horizon every night with northern hemisphere observers getting the best views. The planet is slowly drawing closer to Earth and is a steady magnitude −4.0 this month. It appears next to Jupiter on 2 March when the two planets are less than a degree apart. Venus slices through the plane of the ecliptic on 14 March when it reaches its ascending node and it is occulted by the waxing crescent Moon ten days later on 24 March; this event is visible in south east Asia from sunset onwards. On 30 March, Venus and Uranus are 1.2° apart but a telescope will be necessary to see Uranus in evening twilight.

EARTH reaches its first of two equinoxes on 20 March, ushering in spring to the northern hemisphere and autumn to the south. The Moon continues to block out the planets from time to time, with both Jupiter and Venus falling victim this month.

MARS is visible in the evening sky and is particularly well-placed for observing from northern latitudes. It drops in magnitude from +0.4 to +1.0 as it gets farther from Earth. With east quadrature occurring on 16 March, the red planet looks distinctly gibbous in a telescope, the only superior planet to show such obvious phases. Toward the end of the month, Mars leaves Taurus to amble through the bright stars of the constellation Gemini. The Moon does not occult Mars this month but the waxing crescent Moon does pass 2.3° north of the planet on 28 March. On the same day, Mars has a rather closer flyby of M35, appearing 1.2° north of the open cluster. The stars are scattered over an area about the size of the Full Moon and should not be confused with the nearby compact open cluster NGC 2158.

1 CERES arrives at opposition in ecliptic longitude on 21 March which coincides with its opposition in right ascension and its maximum elongation of 160° from the Sun. The largest member of the main asteroid belt shines at seventh magnitude in the constellation of Coma Berenices. The article *Minor Planets in 2023* has more information.

JUPITER teams up with the evening star low in the west on 2 March, with Jupiter appearing two magnitudes fainter than Venus. Both objects are located in Pisces. Another lunar occultation takes place this month, with Jupiter disappearing behind a sliver-thin crescent Moon on 22 March. This event is visible from sunset in north eastern South America. The largest planet in the solar system is found 1.3° north of the smallest one on 28 March but with conjunction less than two weeks away, even bright Jupiter may be difficult to spot in evening twilight.

SATURN is visible in morning sky and best viewed from the southern hemisphere where it quickly rises before dawn. These observers have the best opportunity to see first-magnitude Saturn and a much-brighter Mercury half a degree apart in the east on the second day of the month. Look for the ringed planet ahead of sunrise in the constellation of Aquarius.

URANUS is now low in the west after twilight as it approaches conjunction in May. Found in Aries, the sixth-magnitude object is only just visible to the naked eye under dark skies so is best observed on moonless nights early in the month. The waxing crescent Moon passes 1.5° north of the planet on 25 March; the year-long cycle of regular lunar occultations finished last month. The evening star is found just 1.2° south of Uranus on the penultimate day of the month.

NEPTUNE is at conjunction with the Sun on 15 March and is not visible this month. Its move from Aquarius to Pisces early in the month, plus its encounter with Mercury and the New Moon on 16 March and 21 March respectively, will go unobserved.

The Incomparable Sir Patrick Moore

Neil Haggath

This year sees the centenary of the birth of the man whose name was associated with the *Yearbook of Astronomy* for almost half a century – Sir Patrick Moore CBE HonFRS FRAS (1923–2012). He was best known, of course, as the presenter of *The Sky at Night*, the longest continuously running TV series in the world, but he was also a highly accomplished amateur astronomer and a remarkably prolific author.

Patrick Alfred Caldwell-Moore was born in Pinner, Middlesex, on 4 March 1923. His love of astronomy began at the age of six, and he joined the British Astronomical Association at 11. He was elected Fellow of the Royal Astronomical Society in 1945.

Patrick Moore standing alongside his 12.5 inch reflecting telescope (nicknamed 'Oscar') in or around 1975. Acquired in late 1940s – and housed in a split run-off shed – this telescope was used by Patrick to map the Moon and later on for lunar and planetary observation. (Brian Jones)

He served in the RAF during the Second World War, was commissioned as a pilot officer, and flew as a bomber navigator. After the War, he became a teacher until 1953. During this time, he built an observatory with a 12.5-inch reflector, which he used for the rest of his life.

He was especially interested in studying the Moon. In 1946, he claimed, together with Hugh Percy Wilkins (1896–1960), to have discovered the difficult to observe Mare Orientale – but later conceded that a little known German astronomer, Julius Franz, had discovered it 40 years earlier.

Moore wrote his first book, *Guide to the Moon*, in 1952. He wrote over 70 books, and also translated several works from French to English, including Eugenios Antoniadi's classic *La Planète Mars* and *La Planète Mercure*. He was Editor of the *Yearbook of Astronomy* for 47 years, from 1965 until his death. When he died in December 2012, the 2013 edition was already on sale, and much of the 2014 edition had already been prepared; the latter was still published in his name, as a Special Memorial Edition.

His television career began with *The Sky at Night* in April 1957; he was fond of saying that the series was older than the Space Age. He presented it for 55 years until his death, missing only one programme due to illness in 2004. For the last eight years, when health problems prevented him travelling, it was presented from his home. In the early days it was broadcast live, and there were a few mishaps. On one occasion, while speaking live on camera, a large fly flew into his mouth; he swallowed it and continued uninterrupted. He delighted in telling how, when he told his mother about this, she said, "Well, it may have been bad for you, dear, but so much worse for the fly!"

He was also a regular presenter during the BBC's coverage of the Apollo missions. He became an instantly recognisable celebrity, and made a number of self-parodying appearances in comedy shows with Morecambe and Wise and The Goodies.

In 1965, Moore was appointed Director of the new Armagh Planetarium in Northern Ireland, but resigned three years later, as he wanted nothing to do with the so-called "Troubles". Returning to England, he settled in Selsey, Sussex, where he lived for the rest of his life.

Moore was an avid traveller, and visited every continent, including Antarctica. He observed several total solar eclipses, and wrote in his autobiography, "There is nothing in nature to match the glory of a total eclipse of the Sun" – a sentiment with which I wholeheartedly agree!

During his career, he met many of the great names of twentieth century astronomy, including Eugenios Antoniadi, Ejnar Hertzsprung, Sir Arthur

Patrick Moore in full flow, addressing an eclipse tour group in Los Angeles, 1991. (Neil Haggath)

Eddington, Harlow Shapley, Edwin Hubble and Gerard Kuiper. He also believed himself to be the only person to have met Orville Wright, Yuri Gagarin and Neil Armstrong!

In 1982, Moore compiled his Caldwell Catalogue of deep sky objects, which was published in *Sky and Telescope* in 1995. As is well known, Charles Messier (1730–1817) compiled his famous catalogue, not due to an interest in deep sky objects, but simply to avoid mistaking them for comets; it therefore omits many of the most prominent objects, as well as those in the southern sky which are not visible from Europe. Moore's Catalogue lists 109 well known objects – deliberately the same number – which are not included in Messier's. He named it the Caldwell, rather than Moore, Catalogue, as the initial M is used for the Messier objects; the objects in it are designated C1 to C109.

He was also a talented musician – a pianist and an accomplished xylophone player – and composed a number of classical works.

Moore was awarded the OBE in 1968, and promoted to CBE in 1988. He was knighted in 2001, for "services to the popularization of science and to broadcasting". Also in 2001, he was appointed Honorary Fellow of the Royal Society – the only amateur astronomer to be so honoured.

Sir Patrick died at his home in Selsey on 9 December 2012, aged 89.

I had the honour of meeting Sir Patrick on several occasions. One of those was in 1991, on my first total solar eclipse trip. While staying in Los Angeles, before different groups headed to Mexico and Hawai'i, we visited Mount Wilson Observatory, where we experienced a minor earth tremor. Patrick's talk to the group that evening included a fine example of his self-mocking humour; referring to the tremor, he said, "Contrary to the rumour, it wasn't caused by me jumping up and down!"

April

Full Moon: 6 April
Last Quarter: 13 April
New Moon: 20 April
First Quarter: 27 April

MERCURY continues its ascent into the evening sky, best seen from northern latitudes. It reaches greatest elongation east (19.5°) on 11 April, after which it heads back toward the horizon. Ten days later, on 21 April, the very young crescent Moon is less than 2° south of the second-magnitude planet but Mercury is very low in the west by this time. On the same day, Mercury enters into retrograde motion.

Evening Apparition of Mercury
17 March to 1 May

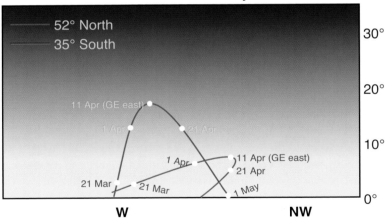

VENUS brightens from magnitude −4.0 to −4.1 this month. Visible in the west after sunset, it is getting a little higher above the horizon every night. Venus finds itself 2.5° south of the open star cluster M45 (the Pleiades) on 11 April. Perihelion occurs six days later when the bright planet is only 0.72 au from the Sun. The waxing crescent Moon draws near on 23 April but is not close enough to occult the planet this month.

EARTH intercepts the debris stream of comet C/1861 G1 Thatcher this month, resulting in the Lyrid meteor shower which peaks just after New Moon. The article *Meteor Showers in 2023* has more details. The lunar occultation of Jupiter on 19 April takes place too close to the Sun to observe but Earth is treated to a rare hybrid solar eclipse the following day. Information on where the eclipse may be seen is found in *Eclipses in 2023*.

MARS continues to inhabit the evening sky. It sets mid-evening from the southern hemisphere but lasts until midnight for observers farther north. The magnitude of the red planet continues to drop as Mars falls behind Earth, starting the month at +1.0 and ending at +1.3. It still appears slightly gibbous in the telescope but is getting ever smaller. Now traversing Gemini, Mars passes just 0.2° north of Mebsuta, the epsilon star of the constellation and two magnitudes fainter than the planet despite being a G-type supergiant. The rocky planet passes another named star of Gemini, Wasat (δ Geminorum), on the last day of the month when the two celestial bodies are 1.9° apart. Wasat shines at around half a magnitude fainter than Mebsuta and is actually a triple star system.

JUPITER is largely lost to view this month, undergoing conjunction with the Sun on 11 April. The lunar occultation that takes place on 19 April is too close to the Sun to be observed.

SATURN continues to rise ever earlier ahead of the Sun but remains mired in dawn twilight as viewed from northern temperate latitudes. The constellation of Aquarius is much better placed for observing from the southern hemisphere; first-magnitude Saturn may be seen in dark skies there. *A finder chart showing the position of Saturn throughout 2023 follows the article The Planets in 2023.*

URANUS is lost in evening twilight fairly early in the month. A very young crescent Moon is only 1.7° north of the planet on 21 April but both objects will be very deep in the glow of sunset.

NEPTUNE is past solar conjunction and has moved into the morning sky, rising at dawn when viewed from northern temperate latitudes but visible in dark skies ahead of the Sun when observed from the southern hemisphere. Now located in the constellation of Pisces, Neptune is passed by the waning crescent Moon on 17 April. At eighth magnitude, the blue ice giant must be viewed through a telescope and is best observed when the Moon is absent from the sky.

Biela's Comet
Life After Death?

Neil Norman

As we discovered in the *Yearbook of Astronomy 2022*, comet 3D/Biela was observed to have returned in two pieces at its perihelion passage of 1846, the twin comets having been last seen in 1852. During the following return of 1859 the comet was badly placed for viewing from Earth, and although the 1866 return was predicted to have the comet ideally placed for observation, it was not recovered. Little did astronomers know that things were to get very interesting indeed.

We now know that meteor showers and comets are connected, due to the debris comets leave behind following repeated perihelion passages around the Sun.

Fig. 287. — La grande pluie d'étoiles filantes du 27 novembre 1872.

The Andromedids of 27 November 1872, as depicted in *Astronomie Populaire* by Camille Flammarion, Paris, 1880. (From the collection of Richard Sanderson)

However, in the early nineteenth century, the two phenomena were not known to be related. 3D/Biela was responsible for two of the strongest meteor showers witnessed during the latter part of the nineteenth century, in 1872 and 1885, in a shower called the Andromedids.

The Andromedids were first recorded in the skies above St Petersburg, Russia, on 6 December 1741, with other strong displays being seen in 1798, 1825, 1830, 1838 and 1847. It is interesting to note that the shower was not overly strong from 1741 until 1758 and that from 1798 until 1847 repeated high rates of meteors were seen. This may suggest that the comet was undergoing stress within its nucleus as early as 1798.

Although the 1866 return eluded observers, plans were in place to search for the comet again in 1872. Astronomers began looking for the comet from around the middle of 1872, and with a perihelion passage due on 30 September of that year hopes were high of finding it, although the comet evaded observers. The searches continued into November but astronomers knew the comet, if it still existed, would be dimming rapidly as it retreated from the Sun.

However, on the night of 27 November things took an unexpected turn when a meteor shower of intense ferocity was seen, with meteors numbering 10,000 per hour raining down from the skies. The German astronomer Johann Friedrich Julius Schmidt, director of the National Observatory of Athens, Greece, noted that the shower consisted mainly of faint fifth and sixth magnitude meteors with a predominantly orange or reddish colouration and with "broad and smoke-like" trains. The display was seen by the English astronomer Edward Joseph Lowe, who claimed to have seen at least 58,600 meteors between 5.50 and 10.30pm.

A similar shower was seen in 1885, during which display the Austro-Hungarian astronomer Ladislaus Weinek obtained the first known photograph of a meteor. The display of 1899 was also strong, although a little less intense.

Were these extreme showers the direct result of the disintegrating comet? Research has indicated that the dust from the initial break up of the nucleus – which seems to

Ladislaus Weinek is credited as having taken the first known photograph of a meteor, this from the Klementinum observatory in Prague on 27 November 1885 (Wikimedia Commons)

have occurred between 1839 and 1844 – would have been dispersed along the comet's orbit by the pressure of solar radiation and placed at the aphelion point. We can deduce from this that the Earth has never encountered the densest material left behind by the dying comet. The dust we encountered that night in 1872 was from the disruptive events of 1846 and 1852. The particles from these returns did not have time to be dispersed along the orbit, thereby allowing the Earth to meet them head-on resulting in a grand display of meteors. With this all in mind, can you imagine if the Earth encountered the densest part of the stream in 1872? It would easily have been 100 times greater.

Sadly today the Andromedid shower that runs between 25 September and 6 December only produces three meteors per hour at best on the peak night of 9 November. This is due to a couple of factors; one, the comet now appears to have completely disintegrated and two, the planet Jupiter has perturbed the orbit of the stream away from the Earth to the point that meteors are now negligible. That said, a moderate display was seen on 5 December 2011, with some 50 meteors being recorded per hour over Canada.

2023 may see another good display with perhaps up to 200 meteors per hour seen. The reason for this is 3D/Biela (should anything remain at all) will come to perihelion on 5 December. It is interesting to note a return would have occurred in 2010, a year before the last enhanced display. Maybe some small fragments of the comet still remain?

What remains of 3D/Biela we simply do not know. Many searches have been conducted in the past, the most notable being that of 1971, when Dr Brian Marsden proposed that if anything of the comet remained it would have expended all of its volatile materials by 1859. To date no asteroidal object of any great size has been discovered to occupy the orbital path of 3D/Biela, although with modern technology becoming far more advanced we may eventually detect the elusive ghost of this comet.

May

Full Moon: 5 May
Last Quarter: 12 May
New Moon: 19 May
First Quarter: 27 May

MERCURY is at inferior conjunction as the month opens and reaches the descending node in its orbit about the Sun three days later. The planet, which is brightening throughout the month, is a morning sky object; this is the best morning apparition of the year for southern hemisphere observers. Mercury reaches aphelion on 14 May and returns to direct motion at around the same time. A greatest elongation west of 24.9° occurs on 29 May.

Morning Apparition of Mercury
1 May to 1 July

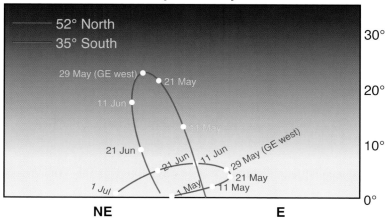

VENUS is finally less than 1 au from Earth. When viewed through a telescope, the evening star is in a gibbous phase and appears a little larger every night, increasing in apparent diameter from 17.1 arc-seconds to 22.8 arc-seconds, and brightening from magnitude −4.1 to −4.2. For observers in northern temperate latitudes, the evening star begins to descend back toward the western horizon. For those farther south, however, Venus continues to climb higher in the sky after sunset. The waxing crescent Moon passes 2.2° north of the planet on 23 May.

EARTH witnesses a faint penumbral eclipse on the fifth day of the month; see *Eclipses in 2023* for details. The Eta Aquariid meteor shower peaks about the same time but will be washed out by strong moonlight. More information may be found in *Meteor Showers in 2023*. Only one lunar occultation of a planet takes place in May and Jupiter is the body taking part.

MARS lingers in the evening sky, best observed from northern latitudes where it sets around midnight. Starting the month in Gemini, the red planet moves into the adjacent constellation of Cancer on 26 May. Mars reaches aphelion on the penultimate day of the month.

JUPITER now appears in the east ahead of the Sun. Southern hemisphere planet chasers have much the best views of this bright object, with astronomers in northern temperate latitudes frustrated by low altitudes and bright dawn skies. On 8 May, Jupiter passes 0.6° north of the fourth-magnitude star Torcular (o Piscium). Nine days later, on 17 May, early risers in much of Mexico and the far west coast of North America may see Jupiter occulted by the waning crescent Moon shortly before sunrise. This event, the final lunar occultation of Jupiter this year, begins at around 11:00 UT. The gas giant departs Pisces for the constellation of Aries the following day.

SATURN is found in Aquarius in the morning sky, reaching west quadrature on 28 May. It is easiest to observe from southern latitudes where it rises around midnight; it is strictly a morning sky object for astronomers farther north.

URANUS is at conjunction with the Sun on 9 May and is largely lost in the glare of the Sun this month.

NEPTUNE continues its slow progress across Pisces in the morning sky. This faint planet is best seen during the latter half of the month from the southern hemisphere where it rises well before the Sun. For observers in northern latitudes, Neptune rises during morning twilight. The waning crescent Moon passes 2.2° south of the planet on 15 May. *A finder chart showing the position of Neptune throughout 2023 follows the article The Planets in 2023.*

Magnetars

John McCue

No, not a 1970s progressive rock band, but a type of neutron star thought to possess an extremely powerful magnetic field – around a hundred million times more powerful than humans have ever made – making it the most magnetic type of star known in the universe. So far, only about thirty magnetars have been found, mostly inside our own Milky Way galaxy.

Turning the clock back on the magnetar's life reveals that it starts out as a massive star, at least eleven times more massive than the sun. The star, like most, produces its energy by converting hydrogen to helium in a controlled way, over a period of many millions of years. We humans can also fuse hydrogen to helium, but in an uncontrolled way with hydrogen bombs – stars are much more sensible!

Over time more massive elements are formed through fusion, until iron – the most stable of all atomic nuclei – is reached. When the star has produced iron, it resists fusion with any other elements. Without this source of fusion energy flowing outwards to balance gravity drawing inwards, the core of the star gives in and begins to shrink. The energy created during this gravitational collapse causes a rise in temperature and pressure and, inevitably, the density rises as well until the electrons are stripped from their nuclei. The result, a white dwarf, is only about the size of the earth – a football made of white dwarf material would weigh about the same as a thousand elephants on Earth.

However, this is only a stepping stone to the magnetar because the star will sidestep the white dwarf scenario if the mass of its core is more than 1.4 solar masses. Gravity now refuses to be resisted and it squeezes the star relentlessly, overcoming the electron pressure. Will the star stop this death crush? It can be done. Some electrons are so energetic with the collapse, that they can penetrate the iron nuclei and combine with the protons therein, crucially forming neutrons. With each reaction a neutrino is also produced, which escapes the star, carrying away its energy. This escape of energy concedes even more to gravity, and so accelerates the rapid decline of the star. Finally, the density does become great enough to stop the march of gravity, but only if the star's core is less than three solar masses. Any more than this and the star has a different fate.

When the collapse suddenly hits the neutron brick wall it rebounds outwards in a shock wave explosion – a supernova. The core itself is now a neutron star, having

The unusual star cluster Terzan 5 in Sagittarius, in which magnetar PSR J1748-2446ad was discovered. Spinning at 716 times per second, this is the fastest magnetar so far known. (NASA/ESA/Hubble/F. Ferraro)

spun faster to conserve angular momentum as it shrank, in just the same way that an ice-skater spins faster when arms are pulled in. The final rotation is incredibly fast – young neutron stars can rotate hundreds of times a second. The fastest pulsar known so far is designated PSR J1748-2446ad. Located nearly 20,000 light years away in the unusual globular cluster Terzan 5 in the constellation of Sagittarius, it spins at 716 times per second (around twice as fast as a typical smoothie blender).

Every star possesses a magnetic field, and when this is concentrated over an area around ten billion times smaller – the neutron star is only the size of a city – the field is likewise ten billion times bigger. By comparison, if the Earth were crushed down to neutron star density, it would be about the size of a football ground. This strong, spinning magnetic field is responsible for the acceleration of charged particles spiralling outward from the neutron star's magnetic poles. This causes the emission of synchrotron radiation from the particles, in the same way that a radio wave is generated here on earth if electrons are pushed up and down an aerial. As the neutron star prodigiously spins, we pick up this radiation as pulses, rather like a flashing lighthouse, across the depths of space, hence its name – pulsar. This synchrotron radiation gradually carries off the rotational energy of the pulsar, and it eventually slows down to a spin every few seconds or so. It is this type of radiation which lights up the Crab Nebula in the constellation of Taurus.

The magnetic field around a magnetar-type pulsar though is about a thousand times that of a 'normal' pulsar. Such magnetism would pull a set of keys out of

An artistic representation of a magnetar – the magnetic field is one hundred million times more powerful than humans have ever produced. (Robert S. Mallozzi, UAH / NASA MSFC)

your pocket at a distance of 200,000 kilometres, equal to roughly half the distance to the Moon. How does such amplified magnetism occur? There are theories. Strong convection currents of charged particles (which are there as well as the neutrons) will be set up in the superfluid interior of the forming neutron star, and will generate a magnetic field in the same way that currents in the liquid iron core of the earth produce our field. If there is a correlation between the rise and fall of the convection currents and the spin of the pulsar, then the overall magnetic field is amplified by around a thousand times. On the other hand, the magnetar could just be the collapse of a star that originally had an unusually high magnetic field. The former theory though is currently preferred because a very strong initial magnetic field would probably act as a brake on the rotation when the star collapsed to its pulsar final stage.

In 1992, American astronomers Robert Duncan and Christopher Thompson predicted the existence of magnetars and developed a model of their structure, and in 2003 they were proved right. Duncan, Thompson and astrophysicist Chryssa Kouveliotou were jointly awarded the Bruno Rossi prize for uncovering this astonishing celestial beast. Chryssa was using NASA's Rossi X-ray Timing Explorer (named after Italian particle physicist, Bruno) and the Japanese Advanced Satellite for Cosmology and Astrophysics, and noticed that a neutron star they were studying was slowing down at just the right rate predicted by Duncan and Thompson to account for the phenomenal magnetic field.

The electrically-conducting surface crust and enormous magnetic field are linked very closely together under incredibly dense, stressful conditions of gripping gravity. Sometimes, this material stress will cause a fracture, a 'starquake' in the surface, as the magnetar slows and adjusts its shape to be as perfectly spherical as possible. Even though such movement may be only fractions of a millimetre, the tied magnetic and electric fields will shift as well, releasing prodigious amounts of energy in the form of gamma rays in a fraction of a second. In 2004, a burst of intense X-rays and gamma rays overloaded the detectors of seven orbiting earth satellites, and the culprit was tracked to a starquake on a magnetar, which had burst out, in 0.2 seconds, a hundred times the energy output of our entire Milky Way galaxy in the same space of time.

After around only ten thousand years or so, the magnetar will slow down, suffer the occasional starquake, cool and lose a lot of its magnetism – a quiet finale to a riotously violent life. This time scale is short in comparison to the lifetimes of other, more law-abiding, stars so there is probably a multitude of magnetars out there waiting to be discovered.

Magnetars are a species of neutron star formed from the supernova explosions of massive stars, and they are terrifying!

June

Full Moon: 4 June
Last Quarter: 10 June
New Moon: 18 June
First Quarter: 26 June

MERCURY begins June at magnitude +0.4 in the dawn sky and ends at a blazing −2.3. For early risers in northern temperate latitudes, this is the worst morning apparition of Mercury this year; conversely, it is the best one for those south of the equator. However, look for the bright planet early in the month as it is heading back toward the eastern horizon and will disappear before the end of June. Mercury is 2.7° north of sixth-magnitude Uranus on 4 June. It passes through its ascending node on 23 June and reaches perihelion four days later.

VENUS reaches greatest elongation east on 4 June when it is found 45.5° away from the Sun. It attains theoretical dichotomy, when it is half illuminated, on the same day, but for reasons not yet understood, observational dichotomy usually differs from theory by several days. From now until October, Venus will appear as a crescent in a telescope. The evening star continues to brighten, increasing in apparent magnitude from −4.3 to −4.4. It also gets higher above the western horizon for southern hemisphere observers although it dips back toward the ground for everyone else. On 13 June, Venus is only half a degree north of the open star cluster M44 (Beehive Cluster or Praesepe).

EARTH reaches a solstice on 21 June, bringing summer to the northern hemisphere and winter to the south. On this day the Sun reaches its greatest declination north during the year. The minor Tau Herculid meteor shower peaks around the time of the Last Quarter Moon which may interfere with observations. More information is available in the article *Meteor Showers in 2023*.

MARS is an evening sky object, best seen from the northern hemisphere where it sets in late evening. Southern observers lose it in the west an hour or two earlier. Having moved into the constellation of Cancer late last month, Mars visits the open star cluster M44 on the second day of June. This cluster is visible to the naked eye and is known as the Beehive Cluster or Praesepe, the 'manger'. Three days later, Mars glides 1.3° north of the binary star Asellus Australis (δ Cancri), the southern donkey which is feeding from the manger. Mars is approximately magnitude +1.6 this month, over two magnitudes brighter than this star. The red planet enters the constellation of Leo mid-month. *A finder chart showing the position of Mars throughout the first eight months of 2023 follows the article The Planets in 2023.*

JUPITER is found in the east before sunrise. The darker winter skies of the southern hemisphere favour observations of the giant planet this month. Now found in Aries, Jupiter will remain in this faint constellation for the remainder of the year. The waning crescent Moon passes 1.5° north of the gas giant on 14 June; there are no further lunar occultations of Jupiter this year.

SATURN shows its minimal ring aspect for the year (7.3°) on 13 June; the rings will now open up slightly over the next five months. The planet enters into retrograde motion mid-month, reversing through the constellation of Aquarius. Saturn finally appears before midnight for northern temperate observers but remains most easily observed from the southern hemisphere where the ecliptic is high overhead in the sky.

URANUS is located in Aries which is a morning constellation at this time of year. It is best viewed from the dark winter skies of the southern hemisphere as the planet remains hidden in the light of dawn at more northerly latitudes. Uranus appears 2.7° south of Mercury on the fourth day of the month and 2.0° south of the waning crescent Moon on 15 June.

NEPTUNE rises very late in the evening as seen from the southern hemisphere but astronomers in northern temperate latitudes must wait until midnight or later for the eighth-magnitude planet to appear in the east in the constellation of Pisces. The waning crescent Moon, just a day past Last Quarter, passes 2.0° south of Neptune on 11 June. Eight days later, on 19 June, the faint planet reaches west quadrature. Retrograde motion in ecliptic longitude commences on 30 June with retrograde motion in right ascension beginning the following day.

Passion for the Stars
The Life and Legacy of William Tyler Olcott

Richard H. Sanderson

Any amateur astronomer with a bit of passion and observing experience can contribute to the advancement of the science. Members of the American Association of Variable Star Observers (AAVSO), a worldwide organization based in Cambridge, Massachusetts, add more than a million variable star observations per year to a database that is utilized by researchers. Comprised mostly of amateur astronomers, the AAVSO was set in motion over a century ago by an energetic stargazer named William Tyler Olcott.

In 1905, Olcott was a 32-year-old non-practicing lawyer living in Norwich, Connecticut, whose financial security had been assured by a bequest from his maternal grandmother. That year, when a friend introduced him to the constellations,

William Tyler Olcott poses with his 3-inch refractor in this picture taken around 1910. (AAVSO)

the flame of passion was ignited. Olcott eagerly learned the night sky and before long, he had gained enough confidence to write a guidebook for beginners.

"The constellations to the vast majority, are utterly unknown," wrote Olcott as he set about to correct that deficiency with *A Field Book of the Stars*. Published in 1907, this slender book focuses on naked-eye astronomy and features constellation maps paired with brief descriptions. Each seasonal chapter begins at the Big Dipper or Pleiades and then jumps from one constellation to the next in a predetermined sequence. *In Starland With a Three-Inch Telescope*, a companion volume for amateur astronomers with small telescopes, appeared two years later.

During the final days of 1909, Olcott travelled to eastern Massachusetts to attend the 61st meeting of the American Association for the Advancement of Science. One of the venues, Harvard College Observatory (HCO), was renowned for its work on stellar photometry carried out under the leadership of Director Edward C. Pickering. This data was put to good use by a small group of variable star observers who routinely sent brightness estimates to HCO. These estimates involved comparing the variables to nearby stars of known magnitude.

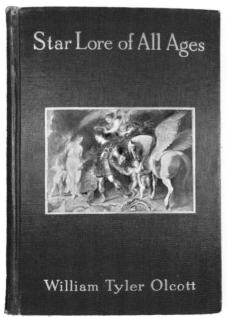 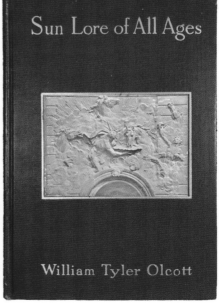

W. T. Olcott's passion for observing was matched by his fascination with the lore and mythology of the heavens. (From the author's collection)

At the meeting, Olcott attended a lecture delivered by Pickering in which he encouraged amateur astronomers to add variable stars to their observing repertoires. Olcott must have been surprised to learn that amateurs with small instruments could make valuable contributions to astronomy. He soon reached out to Harvard College Observatory for more information. Pickering responded by sending HCO night assistant Leon Campbell to provide the necessary instruction. Olcott began making variable star observations in February 1910.

An ambitious new book by Olcott, *Star Lore of All Ages*, was published in 1911. He had endeavoured to "… include all matter pertinent to the subject, in order that the history of the constellations … might be arranged in convenient form for the benefit of those who only know the stars by sight." Olcott gleaned his information from a multitude of astronomy texts, but seems to have been especially captivated by Richard Hinckley Allen's classic 1899 work, *Star-Names and Their Meanings*.

In the city where William Tyler Olcott was born, there lived a young stargazer named Frederick C. Leonard whose energetic enthusiasm had earned him the nickname "Chicago's Boy Astronomer." He and another teenager had created the Society for Practical Astronomy in 1909. Like the British Astronomical

During the AAVSO's first meeting in the Boston area, participants sent this postcard, dated 21 November 1915, to a member who was unable to attend. Among those who signed the card are W. T. Olcott, Edward Pickering, Leon Campbell and several future AAVSO presidents. (From the author's collection)

Association, this new society featured specialized sections, including one devoted to variable stars.

Olcott was eager to share his new-found enthusiasm for variable stars. He penned an article for the March 1911 issue of *Popular Astronomy* and found a valuable supporter in the magazine's editor, Herbert C. Wilson. Olcott would lead the variable star section of the Society for Practical Astronomy for a few years, but before 1911 was over, he announced the formation of an association of variable star observers.

With continued guidance from Pickering and encouragement from Wilson, Olcott grew the AAVSO by recruiting members, creating observing charts and submitting observations to *Popular Astronomy*. He understood the value of periodically bringing members together to expand their knowledge and promote fellowship. In 1915, the AAVSO gathered in the Boston area for the first time. They met Pickering and Campbell, toured Harvard College Observatory and attended the Harvard-Yale football game.

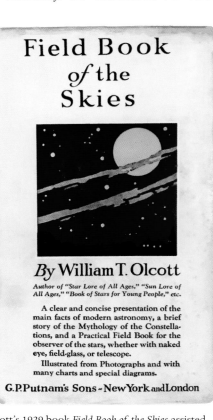

Olcott expanded his survey of celestial folklore and mythology to include our nearest star in his 1914 book, *Sun Lore of All Ages*. With the AAVSO's governance formalized by the association's incorporation in 1918, Olcott took a well-deserved hiatus from book-writing and administrative work and devoted some time to travelling and building an observatory atop his home. Nine years would pass between *Sun Lore* and his next work, *The Book of the Stars for Young People*.

Olcott's literary tour de force came in 1929. *Field Book of the Skies* combines elements from his previous works into a popular guidebook for naked-eye, binocular and telescopic

Olcott's 1929 book *Field Book of the Skies* assisted and inspired generations of stargazers. (From the author's collection)

observers. Within its pages, Olcott proudly announces that the AAVSO's 350 members had contributed 250,000 observations, calling it a "… record of amateur endeavour and achievement unequalled in any other branch of scientific research work."

In later years, Olcott's physical ailments forced him and his wife Clara to spend winters in warmer climes. But it was during a summer evening in 1936, while he was introducing a group of people in New Hampshire to the night sky, that a heart attack claimed his life.

The AAVSO has flourished for more than a century, thanks to the strong foundation created by William Tyler Olcott. He also shared his abiding love of astronomy with the world through his popular books. This rich astronomical legacy owes its existence to a magical evening midway through Olcott's life when a friend opened his eyes to the beauty of the starry heavens.

The author wishes to express his indebtedness to AAVSO Executive Director and CEO Stella Kafka, Ph.D., for providing a photo of William Tyler Olcott, and to Thomas R. Williams and Michael Saladyga, whose 2011 book *Advancing Variable Star Astronomy* masterfully documents the history of the AAVSO.

July

Full Moon: 3 July
Last Quarter: 10 July
New Moon: 17 July
First Quarter: 25 July

MERCURY is at superior conjunction on the first day of the month, appearing in the west after sunset soon after. This is the finest evening apparition of the year for planet watchers in the southern hemisphere and they will have the best opportunity to see Mercury 0.2° north of the Beehive Cluster (M44) on 14 July. Mercury and the evening star have a distant conjunction on 27 July. The following day, the closest planet to the Sun is only 0.1° south of the first-magnitude star Regulus (α Leonis). Mercury passes through its descending node on the last day of the month. This tiny planet begins the month at a bright magnitude −2.3 and ends at +0.1.

Evening Apparition of Mercury
1 July to 6 September

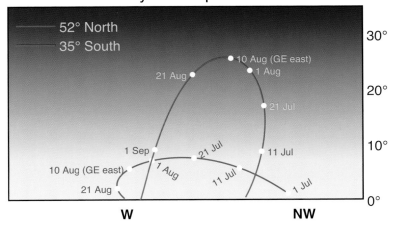

VENUS reaches a maximum brightness of −4.5 this month and rules the western sky after sunset. The bright planet is descending back toward the horizon, rapidly for northern planet watchers, and observers in northern temperate zones will lose sight of the evening star well before the end of the month. However, it lingers awhile longer for those in equatorial and southern regions, not disappearing until the beginning of August. Venus reaches its descending node on 4 July, moving north to south through the ecliptic plane. On 16 July, the bright planet is 1.7° south of Regulus. Retrograde motion commences a few days later. Mercury and Venus have their only conjunction this year on 27 July but they are over five degrees apart at the time.

EARTH is at aphelion on 6 July, 1.017 au from the Sun. The waxing gibbous Moon may inconvenience observations of the Delta Aquariid meteor shower near the end of the month. For more information on this shower, see *Meteor Showers in 2023*.

MARS moves across the constellation of Leo this month, visible toward the west after darkness falls. It passes 0.6° north of Regulus on 10 July, with the star (magnitude +1.4) just brighter than the planet (magnitude +1.7). Fourth-magnitude Rho (ρ) Leonis, a B-type supergiant, is similarly visited by Mars on 21 July when the red planet is found just 0.9° north of the blue star. Mars is starting to become a more challenging object to observe from the northern hemisphere as it sinks into the evening twilight and approaches conjunction later this year.

15 EUNOMIA is an eighth-magnitude object in Sagittarius and passes through opposition in ecliptic longitude on 7 July. Opposition in right ascension and greatest elongation from the Sun (177°) occur hours later on the following day. See *Minor Planets in 2023* for further details.

JUPITER rises before midnight by the end of the month but is still primarily a morning sky object. Found in Aries, the magnitude −2.2 planet is passed by the waning crescent Moon on 11 July, with the two objects 2.2° apart in the sky. *A finder chart showing the position of Jupiter throughout 2023 follows the article The Planets in 2023.*

SATURN is overtaken by the waning gibbous Moon on 7 July with the two bodies 2.7° apart in the constellation of Aquarius. Shining at magnitude +0.7, the ringed planet rises before midnight for all observers but is best viewed from the southern hemisphere.

URANUS appears in the east around midnight this month. Located in Aries, this faint planet requires dark skies in order to see it with the naked eye; southern hemisphere observers should choose a moonless morning during the latter part of the month. The waning crescent Moon is found $2.3°$ north of Uranus on 12 July.

NEPTUNE appears low in the east mid- to late evening. The waning gibbous Moon is $1.7°$ south of Neptune on 8 July but the eighth-magnitude planet is best observed during moonless nights during the latter part of the month. Look for the blue ice giant in Pisces after midnight when the ecliptic is well above the horizon.

134340 PLUTO is at opposition this month near the border between the Capricornus and Sagittarius. Both oppositions, ecliptic longitude and right ascension, take place (about 12 hours apart) on 22 July, as does its greatest elongation of $177°$ from the Sun. This dwarf planet is only fifteenth magnitude and requires powerful optical aids to see it.

A Man (and a Woman) on the Moon

John McCue

The world had just turned a corner, and helping it on its new way was a man whose beliefs would secure this new direction. Theophilus of Alexandria was born around 345 AD[1] in Memphis, Lower Egypt, shortly after the Roman Empire had officially adopted the Christian faith, with far-reaching results for humankind. Theophilus was still young when he lost his Christian parents, leaving him and his younger sister orphaned and in the care of the family's Ethiopian slave. This pious woman took the two children one morning to a Roman temple. Instead of the quiet contemplation they sought, they witnessed the terrifying collapse of the pagan gods' statues – a curious pointer to Theophilus' later life – and moved away for safety to the great metropolis of Alexandria. There, Bishop Athanasius took the three

A sketch of Theophilus of Alexandria based on the fragmented cover of a fifth century papyrus codex. (John McCue)

under his wing when they attended church, and he baptised them. Theophilus was thenceforth educated as a Christian, and his sister as a nun. She eventually married and gave birth to Cyril, who later succeeded Theophilus as bishop.

Theophilus became bishop of Alexandria around the age of forty and is chiefly remembered for his battles against the pagan population of Alexandria. No reliable likenesses of him are known, although the earliest depiction of him appears on the fragmented cover of the Alexandrian World Chronicle, a fifth century papyrus codex. The illustration in question shows the man standing on the ruins of the Roman Serapeum of Alexandria, the great pagan temple and library destroyed by him in 391 AD. The accompanying sketch by the author attempts to bring this

---

1. *Theophilus of Alexandria*, Norman Russell, Routledge, 2007.

cartoon fragmentation to life. Exact events surrounding this momentous obliteration are in doubt, but what is known is that crowd violence between pagans and Christians led to the former taking refuge in the Serapeum. To resolve the standoff, Theophilus sought the Roman emperor's advice, who decided that the pagans should be allowed free, but their serapeum destroyed.

Out now into space, and dominating a promontory on the moon between the Mare Nectaris and the Mare Tranquillitatis is an attractive crater, one hundred and ten kilometres in diameter. Named in Theophilus' memory, it has a fascinating central mountain which divides into three separate peaks and rises three kilometres above the floor. The peak of a complex

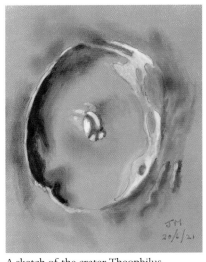

A sketch of the crater Theophilus. (John McCue)

crater such as this indicates the site of the original impact, rebounding material forming this central feature. The walls of Theophilus are terraced, a slumping collapse after the collision, which is again a feature of complex craters. These details are revealed in this sketch by the author of the Theophilus crater, using his own photographic image and sketch details at the eyepiece. Later, enjoying indoor comfort, the picture was executed by scraping grey pastel onto the paper, then rubbing it as a background, followed by the highlights and shadows with pastels, producing the artistic rendition. Immediately to the south west of Theophilus (not shown on the sketch) is the slightly smaller crater Cyrillus, part of the wall of which was destroyed by the later Theophilus-forming collision. This crater was named in honour of Cyril of Alexandria – the nephew of Theophilus – his name having been Latinised to Cyrillus.

Devastated by the destruction of the Serapeum was Hypatia, an Alexandrian astronomer, philosopher and mathematician and contemporary of Theophilus. (Her crater is smaller, of an elongated shape, and is situated in the same region as Theophilus). She was neither Christian nor pagan, but followed Neo-Platonism, believing in the salvation of the soul by studying the beauty and harmony of perfect forms in the natural world. Theophilus thus tolerated her scientific work, so she was untouched by the turbulent times. When Theophilus died in 412 AD however, and was succeeded by Cyril, the mood in Alexandria deepened. He was generally held in poor regard compared with his uncle, and so Hypatia's support was now weaker. Cyril gradually usurped more authority from the Alexandrian prefect, Orestes – a state versus church conflict – and in spite of Orestes and Cyril sharing the same beliefs, they became enemies. Mistakenly thinking that Hypatia was exerting "magical enchanting" over Orestes to prevent his cooperation, Cyril sent his supporters to confront her. Things got out of hand, and Hypatia was brutally murdered.

Forward to the present time again, when the outskirts of the crater Theophilus are chosen as the site for the Japanese SLIM (Smart Lander for Investigation of the Moon) probe, due to launch in January 2022. Its specific target is a small fresh crater just outside Theophilus' walls. This crater shows a strong spectral signature of olivine obtained by the Japanese SELENE orbiter. Olivine is a green mineral common in igneous rocks on earth, and is often polished and used in jewellery but, more importantly, it plays a part in understanding the interiors of planets and moons because it is created there, as are many other minerals, when the body is young and molten. The triple peaks of Theophilus also show signs of internal minerals. How did they get up to the dead Moon's surface? The Japanese astronomers speculate that the olivine – together with other crucial minerals – was blasted to the surface by the mighty smash that created Theophilus and its central peaks. What would the bishop Theophilus of Alexandria have made of this?

A sketch of Hypatia as the author imagines her, after an idea by Jules Maurice Gaspard. (John McCue)

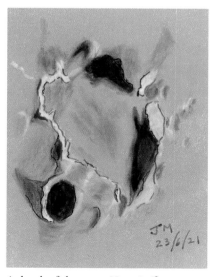

A sketch of the crater Hypatia (from a telescopic image by Steve Doody). (John McCue)

August

Full Moon: 1 August
Last Quarter: 8 August
New Moon: 16 August
First Quarter: 24 August
Full Moon: 31 August

MERCURY is an evening sky object best viewed from the southern hemisphere and equatorial regions where it continues to climb until mid-month before heading back toward the western horizon. The largest greatest elongation east of the year (27.4°) occurs on 10 August, the same day the planet reaches aphelion. Mercury goes into retrograde on 23 August and continues to dim as it heads toward inferior conjunction, ending the month at magnitude +3.2.

VENUS reaches aphelion on 8 August. Its distance from the Sun is 0.73 au which differs little from its perihelion distance (see April) owing to the fact that Venus's orbit is nearly circular. Venus finally disappears from the evening sky for all observers as it undergoes inferior conjunction on 13 August. The nearby planet rapidly reappears in the morning sky and should be 20° above the eastern horizon for most observers by the end of the month.

EARTH has a perfect year for observing the Perseid meteor shower mid-month, with the Moon near its New phase and not drowning the sky in light. *Meteor Showers in 2023* has all of the details. There are also two Full Moons this month. The second is commonly referred to as a 'Blue Moon'. The second Full Moon is also the 'largest' one of the year, that is, it exhibits the biggest apparent angular size on the sky. The most distant apogee of the year takes place on the day of New Moon, 16 August. Finally, the Moon begins a five-year series of occultations of the first-magnitude star Antares (α Scorpii) on 25 August.

MARS finds itself 0.9° south of the fourth-magnitude star Sigma (σ) Leonis on 10 August. A week later, the red planet departs Leo and enters into the constellation of Virgo. The waxing crescent Moon passes 2.2° north of Mars on 19 August. Four days later, on 23 August, Mars scrapes past the beta star of Virgo known as Zavijava. The two objects are about an arc-minute apart at closest approach. Mars remains

Morning Apparition of Venus
August 2023 to June 2024

an evening sky object but is becoming difficult to spot from northern temperate latitudes as it languishes low in western twilight. The dark winter skies of the southern hemisphere may offer better observing opportunities of the second-magnitude planet this month.

JUPITER reaches west quadrature on 7 August, rising late in the evening. On the following day, the Last Quarter Moon passes just under 3° north of the planet. Jupiter narrowly misses occulting the sixth-magnitude star Sigma (σ) Arietis on 21 August. Jupiter passed in front of this naked-eye star in 1952, observations of which led to pre-space-age discoveries about Jupiter's then-mysterious atmosphere. For more information about this event and other stellar occultations by Jupiter, see the article *Shining a Light on Jupiter's Atmosphere* following this month's Sky Notes.

SATURN comes to within 2.5° of the Moon twice this month, on 3 August and again on 30 August. The planet shines at magnitude +0.5 and is visible throughout the night as it reaches opposition in Aquarius on 27 August. When viewed through a telescope, the rings are open to 9°. *A finder chart showing the position of Saturn throughout 2023 follows the article The Planets in 2023.*

URANUS is just 2.6° south of the waning crescent Moon on 9 August and reaches west quadrature a week later. Its trek across the constellation of Aries has been slowing and it reaches a stationary point on 29 August, changing from direct to retrograde motion. Uranus now rises in late evening but is best observed during the morning hours when it is highest in the sky.

NEPTUNE continues to rise earlier every evening and is found in the east in Pisces as twilight ends. The waning gibbous Moon is only 1.5° south of Neptune on the fourth day of August; it has been drawing nearer all year and will actually occult the distant world next month.

Shining a Light on Jupiter's Atmosphere

Lynne Marie Stockman and David Harper

On 7 December 1995, NASA's *Galileo* spacecraft sent a probe into the atmosphere of Jupiter. For almost an hour, the probe's six instruments sent back detailed information about the upper 150 kilometres, answering questions that had puzzled astronomers for more than a century. This was not the first attempt to explore the gas giant's atmosphere. Spectroscopy had been tried as early as 1863, but Jupiter's visible spectrum is dominated by reflected sunlight. In 1903, the Dutch astronomer Anton Pannekoek suggested that occultations of bright stars might offer a way to explore planetary atmospheres, but the technology to carry out the required observations did not exist until after the Second World War.

Pannekoek's idea was put to the test on 20 November 1952, when Jupiter occulted σ Arietis, a magnitude +5.5 blue main sequence star. American astronomers William Baum and Arthur Code observed the event, using the 60-inch telescope at the Mt Wilson Observatory and the then-recent innovations of efficient photomultiplier tubes and a fast recording device. Jupiter was 1,400 times brighter than the star, which presented a challenge, but Baum and Code solved this problem in two ways. First, they restricted the field of view to an 8 arc-second circular region at the limb of the planet adjacent to the star, so that only 1–2% of Jupiter's disk was visible. They also arranged for the photometer to sample a very narrow spectral band around the wavelength of the K line of calcium. The Sun has a very strong K line, and since Jupiter shines by reflected sunlight, it is relatively

NASA's *Juno* spacecraft is the latest visitor to the Jovian system. The orbiter reached Jupiter in July 2016, its mission to study the formation and evolution of the giant planet, observe Jupiter's gravitational and magnetic fields, and to look deep into the Jovian atmosphere. In this image from 29 May 2019, *Juno* captures Jupiter's tumultuous northern hemisphere. (NASA/JPL-Caltech/SwRI/MSSS/Kevin M. Gill © CC BY)

faint at this wavelength. By contrast, σ Arietis is a B-type star which has virtually no K absorption line. This choice of wavelength by Baum and Code enhanced the brightness of the star relative to Jupiter, and together, the two techniques reduced the disparity to just 2:1.

The two astronomers recorded the light curve as the star disappeared behind Jupiter. From this, they were able to estimate the mean molecular weight of Jupiter's outer atmosphere, something that had been calculated only indirectly before. It confirmed that the upper stratosphere of Jupiter was most likely to be hydrogen and helium rather than methane. These results were published in *The Astronomical Journal* in 1953.[1] Their paper has been cited by other astronomers more than 120 times and continues to be referenced today.

Jupiter has occulted seven naked-eye stars since its 1952 encounter with σ Arietis. The 1957 occultation of 13 Virginis was unobservable, so the next opportunity to probe Jupiter's atmosphere came on 13 May 1971 when Jupiter occulted the multiple star β Scorpii. Observatories throughout the southern hemisphere were pressed into service by international teams of astronomers.

In a small telescope, β Scorpii appears as a binary star, historically called β¹ Scorpii and β² Scorpii. Both 'stars' are actually trinary systems but only the brightest component of each triple was observed during the occultation. Instrumentation had improved since the 1950s, and astronomers were able to measure fine details of both stellar immersion and emersion in multiple wavelengths. This included video of the brightest member of β Scorpii 'flashing' several times as it vanished behind Jupiter, and again when it reappeared on the other side. From this data, scientists attempted to construct detailed models of Jupiter's upper atmosphere as well as to devise constraints for Jupiter's radius and oblateness.

By happy coincidence, the Jovian satellite Io also occulted β Scorpii, allowing astronomers to estimate its radius as 1,829 kilometres, which is within 0.5% of the current accepted value. An upper limit for the pressure of the then-unknown Ionian atmosphere was also calculated.

Between 1973 and 1992, five space probes examined Jupiter as they flew past, and in 1995, *Galileo* began its eight-year mission to explore the planet and its major satellites. Scientists now had data directly from the gas giant. Perhaps this is why the next three occultation events – ω Ophiuchi in 1995, ν² Sagittarii in 1996 and HD 201057 in 1997 – went largely unobserved.

Since the end of the *Galileo* mission, two more naked-eye occultations have taken place. Sixth-magnitude 45 Capricorni fell victim to Jupiter on 3 August 2009.

---

1. Baum, W.A., Code, A.D. 1953, 'A photometric observation of the occultation of σ Arietis by Jupiter', *The Astronomical Journal*, **58**, 108–112. **doi.org/10.1086/106829**

Professional and amateur astronomers, working together in a variety of locations and using telescopes ranging in size from 0.4 metres to 2.2 metres, obtained results that were consistent with both earlier occultations and *Galileo* data. This Pro-Am collaboration highlighted the continuing need for ground-based occultation measurements in the intervals between space missions and the importance of having multiple independent observations of the same occultation event. It also demonstrated the ability of amateur astronomers with small telescopes to contribute to serious scientific enquiry.

The most recent bright star occultation took place on 2 April 2021 when sixth-magnitude 44 Capricorni disappeared behind the limb of Jupiter. The Jovian moons Io and Amalthea also occulted the star. This was a particularly difficult occultation to observe, however, as the planet and star were only 49° away from the Sun. The event was observed from Brazil and islands in the Caribbean, although there had been no published results at the time this article was written.

Now, nearly 71 years after its famous occultation, Jupiter once again passes close by σ Arietis. On 21 August 2023 at 21:58 UT, the centre of Jupiter will be just 29 arc-seconds north of the star, with the limb of the planet missing σ Arietis by less than 10 arc-seconds. Closest approach is best seen from locations in time zones approximately two to eight hours ahead of GMT (Eastern Europe and Africa, Asia and Western Australia). Observers in the United Kingdom and Republic of Ireland will still be in nautical twilight with Jupiter just rising in the east. Powerful binoculars or a small telescope will be necessary to see both the star and Jupiter which is eight magnitudes brighter. See the accompanying diagram for more details.

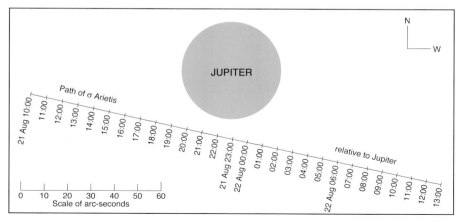

This chart depicts the path of σ Arietis relative to Jupiter during the close approach on 21 August. Jupiter is moving eastwards and passes north of the star. This is the view that would be seen in binoculars. Times are UT. (David Harper)

September

Last Quarter: 6 September
New Moon: 15 September
First Quarter: 22 September
Full Moon: 29 September

MERCURY is not visible early in the month, undergoing inferior conjunction on 6 September. Afterwards it moves into the morning sky in what is the best dawn apparition of the year for observers at northern temperate latitudes. It grows steadily brighter in the east, ending the month at magnitude −1.0. After returning to direct motion mid-month, Mercury goes through its ascending node on 19 September and reaches greatest elongation west (17.9°, the smallest of 2023) three days later on 22 September. Perihelion occurs the following day.

Morning Apparition of Mercury
6 September to 20 October

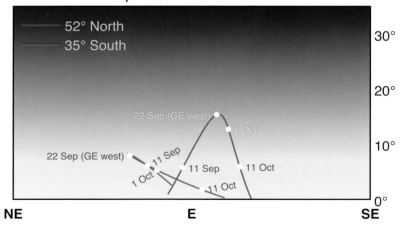

VENUS rules the dawn sky as the morning star, gaining altitude above the eastern horizon every day. Reaching a maximum magnitude of −4.5 by mid-month, Venus appears as a waxing crescent in a telescope. The size of the disk is dwindling, from 50.3 arc-seconds to 32.2 arc-seconds, as the planet draws away from Earth. Direct motion resumes early in the month.

EARTH goes through its second equinox this year on 23 September, with spring returning to the southern hemisphere and autumn beginning in the north. The Moon is busy occulting planets this month, with Neptune disappearing twice and Mars once. On the fifth day of September, the Moon begins a six-year series of occultations of Alcyone (η Tauri), the brightest member of the open star cluster known as the Pleiades (M45). Antares (α Scorpii) is also occulted on 21 September. The Full Moon at the end of the month is known in the northern hemisphere as the 'Harvest Moon', the Full Moon nearest to the September equinox.

MARS may be lost to view for observers in northern temperate latitudes, with the red planet very close to the horizon after sunset. The red planet shines at magnitude +1.8 this month and passes close by two named stars in Virgo, Zaniah (η Virginis, magnitude +3.9) on 4 September and Porrima (γ Virginis, magnitude +2.7) on 12 September. The very young crescent Moon returns to occult Mars on 16 September but observations of this event can only be attempted from a small region of South America including Suriname, French Guinea and northern Brazil. Look for the two objects in the west as the Sun sets.

JUPITER is rising ever earlier in the evening but is best viewed in morning skies when the constellation of Aries is higher in the sky. The bright planet enters into retrograde motion on the fourth day of the month, a sure sign that opposition is approaching.

SATURN is just past opposition and visible for most of the night in Aquarius. The constellation of the Water Carrier never gets very high when seen from northern temperate latitudes; better views are available from the southern hemisphere. The waxing gibbous Moon passes 2.7° south of the planet on 27 September.

URANUS rises earlier every night and appears in the east mid-evening. Located in the faint constellation of Aries, Uranus is only sixth-magnitude and requires dark skies in order to see with the naked eye. The waning gibbous Moon is just 2.8° north of the green orb on 5 September.

NEPTUNE is occulted by the Moon twice this month, on 1 September and again on 28 September. However, these events are only visible from Antarctica and the glow of the nearly Full Moon will heavily interfere with telescopic observations of the planet at these times. Be that as it may, the waning crescent Moon will not pose a problem with observing the sight of Neptune passing just 0.1° north of the fifth-magnitude K-type giant star 20 Piscium on 12 September. As always, a telescope will be necessary to view this appulse since Neptune is only magnitude +7.8 with an apparent disk diameter of 2.53 arc-seconds, well below the naked-eye limit. The blue ice giant is visible all night, coming to opposition on 19 September. *A finder chart showing the position of Neptune throughout 2023 follows the article The Planets in 2023.*

The Letter that Proved the Island Universe Theory

David M. Harland

A century ago, in February 1924, Edwin Hubble wrote a letter to his rival Harlow Shapley, casually explaining a discovery that undermined Shapley's view of the universe.

-o0o-

In 1912 Henrietta Swan Leavitt at the Harvard College Observatory wrote up her pioneering study of variable stars in the Small Magellanic Cloud, reporting the discovery of an empirical relationship between the period of the light curve of a certain type of star (now called Cepheids) and its luminosity.[1]

In 1914 Harlow Shapley joined the Mount Wilson Observatory in the San Gabriel Mountains, northeast of Los Angeles, and set about using the 60-inch reflector, the largest operational telescope in the world, to determine how globular clusters were arranged in 3-dimensional space. To start with, he calibrated Leavitt's relationship so that he could use Cepheids to measure the distances to those globulars in which he could identify such stars. Then he used a 'ladder' of assumptions to estimate the distances to others. Publishing his results for 69 globulars in 1918, he said the globulars were spherically distributed, and that if the Milky Way system of stars had the same centre then it must be a lens-shaped disk about 300,000 light years in diameter. This was far bigger than had been believed and, indeed, was almost inconceivably large.

Meanwhile, astronomers had been discovering a great many spiral nebulae. Most reckoned they were nearby objects, perhaps new stars in the process of forming planets; others suspected they were independent 'island universes' and that if the Milky Way system could be observed from outside, then it too would display a spiral form.

For its 1920 meeting in Washington, D.C., the National Academy of Sciences invited two speakers to draw the evening to a close by offering differing opinions

---

1. I discussed Henrietta Swan Leavitt's discovery in the *Yearbook of Astronomy 2021*.

The 100-inch reflector of the Mount Wilson Observatory. (Observatories of the Carnegie Institution for Science Collection at the Huntington Library, San Marino, California)

on the 'The Scale of the Universe'. Shapley presented the case for the Milky Way system comprising the entire universe, with the implication that the spirals were local. He was followed by Heber Curtis of the Lick Observatory in California, who said the spirals were systems of stars comparable to the Milky Way system. The debate was inconclusive, with neither side conceding defeat. What was required to resolve this debate was incontrovertible evidence one way or the other.

In 1919 Edwin Powell Hubble joined the Mount Wilson Observatory and set about using the unprecedented resolving power of the recently commissioned 100-inch reflector to take sharp photographs of nebulae of various types.

On 4 October 1923, having decided to search for novae in the Great Spiral in Andromeda (M31), Hubble identified a possible nova. The next night, he confirmed the nova and found two additional candidates. On later examining archived plates, he found that one of his three possible novae was actually a variable star. What is more, by plotting its light curve over the ensuing months he determined its period

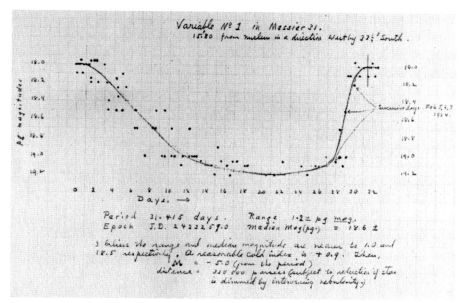

The light curve which Edwin Hubble manually plotted for the first Cepheid variable to be discovered in M31. (Observatories of the Carnegie Institution of Washington)

to be 31.4 days with the characteristic profile of a Cepheid. Leavitt's period-luminosity relationship was about to change our perception of the universe.

In early 1924 Shapley, now director of Harvard College Observatory, received a letter from Hubble dated 19 February which casually stated, "You will be interested to hear that I have found a Cepheid variable in the Andromeda Nebula." Having supplied his hand-drawn light curve, Hubble said, "… rough as it is, (it) shows the Cepheid characteristics in an unmistakable fashion". After describing some other results, he finished off saying, "I have a feeling that more variables will be found by careful examination

Edwin Hubble. (Wikimedia Commons/Johan Hagemeyer)

of long exposures. Altogether next season should be a merry one and will be met with due form and ceremony." The friendly style belied the fact that the two men detested one another.

Discovering a Cepheid in M31 was a major milestone in astronomy. Applying Shapley's calibration of Leavitt's relationship, Hubble found M31 to be a vast system of stars almost a million light-years away. This resolved the Great Debate.

The news spread informally by word of mouth, then by an article in *The New York Times* on 23 November 1924, and formally on 1 January 1925 in a paper at a joint meeting of the American Astronomical Society and the American Association for the Advancement of Science held in Washington, D.C.[2]

Hubble went on to make even more momentous use of the 100-inch telescope, providing observations that revealed the universe to be in a state of expansion, rather than static as had been believed. He died in 1953 at his home in San Marino, California.

Further Reading
The Day We Found the Universe by Marcia Bartusiak, Pantheon Books, 2009.

Note
Shapley's 300,000 light years for the diameter of the Milky Way system and Hubble's million light-years for the distance to M31 are the figures given at the time. After refinements, these distances were respectively almost cut in half and more than doubled.

2. 'Cepheids in Spiral Nebulae', E. P. Hubble, *Publications of the American Astronomical Society*, 5, pp. 261-264, 1925.

October

Last Quarter:	6 October
New Moon:	14 October
First Quarter:	22 October
Full Moon:	28 October

MERCURY begins October visible in the morning sky but it is losing altitude and disappears before mid-month, with superior conjunction taking place on 20 October. Mercury's final evening apparition of 2023 gets underway afterward but it is barely visible from northern temperate latitudes; head to the southern hemisphere to see this tiny planet over the next two months. Mercury's occultation by the Moon on 14 October occurs during daylight hours. It moves through its descending node on 27 October and is found passing 0.3° north of Mars two days later.

VENUS is visible in the morning sky ahead of sunrise. Its rapid ascent into the sky slows substantially this month, although the magnitude −4.5 object is impossible to miss in the east and is best viewed from northern and equatorial regions. Venus appears 2.3° south of Regulus, the lucida of Leo, on 9 October. Theoretical dichotomy occurs on 22 October, the day before greatest elongation west (46.4°). Venus cruises through its ascending node on 25 October, moving north of the plane of the ecliptic where it will remain for the rest of the year.

EARTH and its Moon participate in an annular solar eclipse on 14 October and a partial lunar eclipse two weeks later on 28 October. For observing details, see the article *Eclipses in 2023*. The Moon occults Mercury just before the solar eclipse on 14 October and Mars the following day, but neither event will be visible. The Pleiades (M45) is occulted twice, on 3 October and again on 30 October, with Antares (α Scorpii) disappearing behind the waxing crescent Moon on 18 October. Three meteor showers, the Draconids and the southern branch of the Taurids early in the month, and the Orionids two weeks later, are mostly untroubled by moonlight. The article *Meteor Showers in 2023* explains more.

MARS will be at conjunction with the Sun next month and is largely lost in the bright evening twilight. Its movement past Spica, the brightest star in the constellation Virgo, on the third of the month, as well as Kang (κ Virginis) on 19 October, may occur too close to the horizon to be observable. The final lunar occultation of the red planet takes place on 15 October, just a day past New Moon, and is visible only from the Antarctic. Mars enters Libra late in the month and has a close encounter with Mercury on 29 October, a sure sign that the Sun is nearby.

JUPITER is in retrograde in the constellation of Aries this month, and is best observed from the darker autumnal skies of the northern hemisphere. It rises in early evening and is visible for most of the night. *A finder chart showing the position of Jupiter throughout 2023 follows the article The Planets in 2023.*

SATURN is visible in the evening, already well aloft by the time darkness falls. The ringed planet is found in Aquarius and sets just after midnight. The waxing gibbous Moon passes 2.8° south of first-magnitude Saturn on 24 October.

URANUS has two close encounters with the Moon this month, appearing 2.9° south of our satellite on both 2 October and 30 October. Located in Aries, it rises in the early evening and is well-placed for viewing from both hemispheres.

NEPTUNE is past opposition and already above the horizon once the Sun has set. Located in Pisces, it is well-placed for telescopic observation from both hemispheres. The waxing gibbous Moon passes 1.5° south of Neptune on 26 October but choose a moonless night early in the month to look for the most distant planet in the solar system.

The Diary of a Long Distance Pioneer

Peter Rea

Introduction

This is an imaginary first-person account of the journey of Pioneer 10. For a narrative of the Pioneer 10 mission see 'A True Pioneer of Planetary Exploration' by Neil Haggath in the *Yearbook of Astronomy 2022*.

Dear Diary…

2 March 1972

I feel movement! The rocket that will send me to Jupiter is lifting off. I'm sat on top of an Atlas Centaur rocket with an additional third stage. In the years before launch they called me Pioneer F. I have been called that ever since I was conceived on a drawing board and will retain the letter F until I'm safely on my way to Jupiter.

3 March 1972

What a ride! That additional third stage increased my velocity to 51,682 km/h. This enabled me to cross the orbit of the Moon in just 11 hours, making me the fastest man-made object at that time. During that third stage boost I was spinning at 30 rpm, but I have now slowed down to 4.8 rpm. This spin helps keep my main dish antenna pointing back to Earth. As this launch was successful, I will go down in history as Pioneer 10.[1]

At 01.49 UT on 2 March 1972 Atlas Centaur AC27 launches Pioneer F on a trajectory that would take it past Jupiter and out of the solar system. (NASA)

---

1. It was common practice to give the Pioneer and Mariner missions letters of the alphabet during the design and planning stages. The mission number was only given after a successful launch. Pioneer F became Pioneer 10 and Pioneer G became Pioneer 11.

15 July 1972

It is 136 days since I was launched. I have just entered the region known as the asteroid belt. There is no way round the asteroid belt, and I must travel through it to reach the first of the gas giant planets. I'll keep a lookout for any dangerous rocks. They called me Pioneer for a reason. I'm the first to enter this region.

15 February 1973

It's 351 days since launch and I have just exited the asteroid belt. I spent 216 days within the belt. I expected this region to be crowded with rocks and perhaps dangerous to traverse but the region from my perspective was remarkably void of asteroids. The closest I ever came to one of the asteroids was 8.8 million kilometres. The asteroid belt is clearly no hazard to my future sister spacecraft being sent to the outer planets. So it's onward to Jupiter!

Pioneer 10 image of Jupiter taken on 1 December 1973 at 2,557,000 kilometres. Note how the Great Red Spot is more prominent when Pioneer passed by than when the Voyager probes later visited Jupiter. (NASA)

6 April 1973

Good news! My sister spacecraft Pioneer G was launched successfully today and will now be known as Pioneer 11. Watch out for rocks in the asteroid belt but I think the way is clear.

6 November 1973

Jupiter is getting closer by the hour. Today I am 25,000,000 kilometres from Jupiter and in two days I will cross the orbit of an outer moon called Sinope which at the time of my arrival is the outermost known moon of Jupiter.

29 November 1973

The pace of activities is picking up. I am now inside the orbits of the outer satellites. I look forward to seeing the large moons Io, Europa, Ganymede and Callisto.

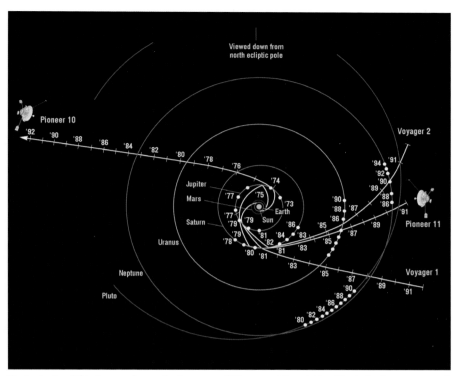

Trajectories. This image shows the journey of Pioneers 10 and 11 with their sister spacecraft Voyagers 1 and 2 as they leave the solar system. (NASA)

3 December 1973

I have just made my closest approach to the four Galilean satellites. Although I came within 321,000 kilometres of Europa my path through the Jovian system put me a bit farther from the other three. I could only get to within 1,392,300 kilometres of Callisto.

4 December 1973

I have just passed within 132,252 kilometres of the cloud tops of Jupiter. All my instruments were active recording the radiation and magnetic fields etc. I managed to take around 500 images of my passage through the Jovian system. My job is almost done. Once I have returned all my data, I will be heading out of the solar system forever. I wonder what I will find.

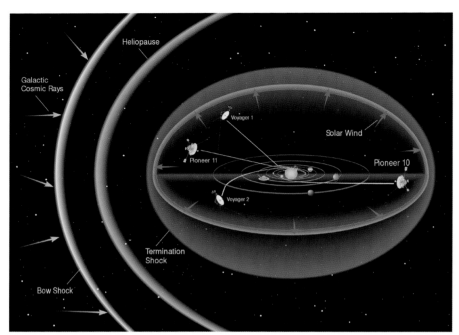

Heliopause (NASA). The paths of the two Pioneers with their sister spacecraft Voyagers 1 & 2 as they reach the edge of our solar system and prepare to cross the heliopause into interstellar space. (NASA)

25 April 1983

Today I crossed the orbit of Pluto. Due to the highly elliptical shape of the orbit, Pluto was closer to the Sun than Neptune when I was out there.[2]

13 June 1983

Today I crossed the orbit of Neptune which at the time was the furthest planet. I had previously crossed the orbit of Saturn in 1976 and Uranus in 1979. The edge of the solar system beckons.

31 March 1997

Today marks the official end of my mission. Controllers will contact me from time to time and I will send whatever data my remaining instruments can send. I hope I did them proud.

---

2. Pluto was still classified as a planet when Pioneer 10 crossed its orbit.

Pioneer Deep Space. An artist's impression of Pioneer 10 in interstellar space. (NASA Ames)

23 January 2003

It has been 11,285 days or 30 years 10 months and 22 days[3] since I was launched. I am heading toward the constellation of Taurus in the general direction of the star Aldebaran. It could take two million years before I get anywhere near. Before that I will pass within 0.75 light years of a K type star in the constellation Cassiopeia. Ahead of me lies the heliopause, a bubble-like region surrounding the solar system caused by the solar wind. It saddens me that I will have no power left to record the effects as I pass through this boundary and enter interstellar space around the year 2020. Power is very low. I cannot last much longer. My controllers still try and contact me. Will I ever be found? I must sleep now never knowing my fate. Eternity is a long time. BLEEP, BLEEP, BLEEP, BLEEP, BLEEP...

Further Reading

Fimmel, Richard O; Van Allen, James; Burgess, Eric. *Pioneer, First to Jupiter, Saturn and Beyond* NASA SP-446 1980

Fimmel, Richard O; Swindell, William; Burgess, Eric. *Pioneer Odyssey* NASA SP-349 1977.

Pioneer Odyssey is available online at **history.nasa.gov/SP-349/sp349.htm**

---

3. Calculated by **timeanddate.com**

November

Last Quarter: 5 November
New Moon: 13 November
First Quarter: 20 November
Full Moon: 27 November

MERCURY is once again found in the west after sunset. It remains very low to the horizon for observers in northern temperate latitudes but is a much easier object for astronomers farther south. It is slowly dimming, from magnitude −0.8 to −0.4 over the course of the month. Mercury reaches its final aphelion of the year on 11 November and is passed by the very young crescent Moon three days later when the two objects are less than 2° apart. The tiny planet is brighter than the first-magnitude star Antares (α Scorpii) which it encounters on 16 November.

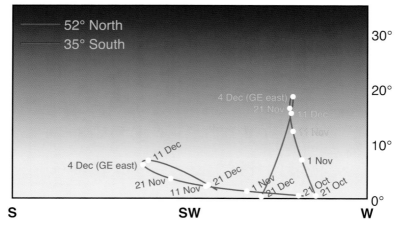

Evening Apparition of Mercury
20 October to 22 December

VENUS is the morning star, slightly declining in altitude when viewed from the northern hemisphere but still climbing as seen from southern latitudes. The waning crescent Moon occults the bright planet on 9 November. This event is visible from

Greenland, Iceland and Svalbard from around 9:00 UT. Venus attains perihelion for the second time this (terrestrial) year on 28 November.

EARTH enjoys several meteor showers this month, including the northern branch of the Taurids as well as the rather more famous Leonids. The article *Meteor Showers in 2023* gives the observing details. The Moon occults only one planet this month, our neighbour Venus, but continues on its monthly path in front of both the first-magnitude star Antares (α Scorpii) on 14 November and the open star cluster M45 (Pleiades) on the day of the Full Moon.

MARS is at conjunction with the Sun on 18 November and is lost to view this month.

JUPITER attains opposition on the third day of month. Shining a brilliant magnitude −2.9, the planet is nearly 50 arc-seconds wide as seen through a telescope. The largest planet in the solar system is visible all night in the constellation of Aries, with the waxing gibbous Moon passing 2.8° north of it on 25 November.

SATURN returns to direct motion early in the month and the rings begin to close up again after reaching a maximum opening of 10.5° on 2 November. The Full Moon passes 2.7° south of the first-magnitude planet on 20 November. Saturn attains east quadrature three days later. The ringed planet is in the constellation of Aquarius and is visible as darkness falls, setting before midnight.

URANUS reaches opposition on 13 November, rising at sunset and setting at sunrise. It is at its brightest, magnitude +5.6, and appears as a green disk 3.68 arc-seconds in diameter in the eyepiece of a telescope. This is an excellent opportunity to see this faint planet as the Moon will be absent from the sky. On 26 November, the waxing gibbous Moon passes 2.7° north of the planet which is located in the constellation of Aries. *A finder chart showing the position of Uranus throughout 2023 follows the article The Planets in 2023.*

NEPTUNE is in the constellation of Pisces at the beginning of November but retrograde motion takes it back to Aquarius before the end of the month. The waxing gibbous Moon passes 1.5° south of the eighth-magnitude planet on 22 November. Neptune is already aloft once the sky turns dark and sets after midnight.

Did We Receive A Radio Signal From Proxima Centauri?

David M. Harland

On 18 December 2020 the British newspaper *The Guardian* reported a 'scoop' that raised the possibility of an artificial radio signal originating from Proxima Centauri, the nearest star to the Sun.

In 2015 the billionaire venture capitalist Yuri Milner provided $100 million to establish Breakthrough Listen, whose purpose is to observe one million of the closest stars for 'technosignatures' as indications of alien existence. As such, it is one of a number of projects undertaken by the SETI (Search for Extraterrestrial Intelligence) community.

Over the decades, the SETI effort has detected numerous 'transients' in the radio spectrum, most notably a strong narrowband signal which was spotted in 1977 by the Big Ear antenna of the Ohio State University Radio Observatory.

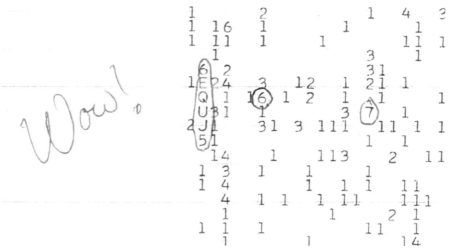

The "Wow!" signal detected on 15 August 1977 at 1,420 MHz (a wavelength of 21 centimetres) at which neutral hydrogen emits. It lasted 72 seconds and came from the direction of the constellation of Sagittarius, where the centre of the galaxy is located. (Big Ear Radio Observatory and North American Astrophysical Observatory)

In the southern sky, the Breakthrough Listen project 'piggybacks' on observations made using the 64-metre dish of the Parkes Observatory in Australia.

While Parkes was conducting a survey of flare activity on red dwarf stars, it made a series of observations of Proxima Centauri starting on 29 April 2019, with five 30-minute sessions over a total of 30 hours on several days running into early May.

When Shane Smith, an undergraduate student at Hillsdale College in Michigan and an intern at Breakthrough Listen, studied the data in late October 2020 he spotted a signal at a narrow frequency of 982.002 MHz that was present during each of the five times the telescope was aimed at Proxima Centauri and absent when it was not.

The project uses software filters which reject the multitude of signals originating from Earth or spacecraft, and this transmission was unlike anything the team had encountered. It was so intriguing that it became the first detection to achieve 'candidate' status, as Breakthrough Listen Candidate (BLC) 1.

When astronomers were alerted to the existence of the signal in the recorded data, they took another look, without result. It has therefore been classified as a transient. If ever the signal reappears, and can be observed simultaneously by widely-separated telescopes this will allow many forms of terrestrial interference to be eliminated.

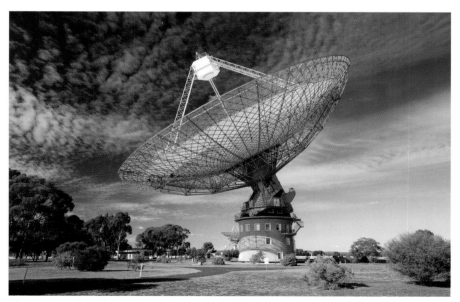

The 64-metre dish of the Parkes Observatory. (CSIRO)

It is not possible to say for sure that the signal originated at Proxima Centauri, 4.2 light years away, only that its point of origin was within a circular patch of sky defined by the beam width of the telescope, which was about half the angular diameter of the lunar disk, centred on the star's position. However, the possibilities are intriguing.

In 2016 a planet was reported to be orbiting Proxima Centauri. The discovery was made by measuring how the planet's presence induced cyclical Doppler shifts in the spectral lines of the star, measured by the High Accuracy Radial velocity Planet Searcher (HARPS) spectrograph on the 3.6-metre ESO telescope at La Silla in Chile for the 'Pale Red Dot' project. In accordance with convention, the first planet to be identified in the system was designated Proxima b.

Being a cool M-type dwarf, the habitable zone of Proxima Centauri is quite close in. The radius of Proxima b's orbit is roughly 0.05 au, so its orbital period is only 11.2 of our days. The inclination of the orbit is not known, but the mass is at least 1.2 times that of Earth. Although it is a rocky world orbiting within the habitable zone of the star, and as such could theoretically have liquid water on its surface, the radiation from the flares that occur on such stars could have made it very difficult for life to develop there. In 2020 HARPS found a second planet. This 'super Earth' orbits with a period of 5.2 years, and is well outside the habitable zone.

Artistic impression of the planet Proxima b orbiting the red dwarf star Proxima Centauri, which is itself a distant companion to a pair of the much brighter stars seen in the distance. (ESO/M. Kornmesser)

In effect the narrow bandwidth of the signal at 982.002 MHz made it a single tone. That said, there was some Doppler shift, but nothing that could be identified with the motion of a planet orbiting the star. There was no sign of modulation (although it is conceivable that information was being conveyed by varying the polarisation). Another speculation is that the monotone signal represented a 'beacon' whose purpose was to say 'look here, more to follow'.

In a tweet shortly after the news broke, Peter Worden, an executive director of Breakthrough Initiatives with a doctorate in astronomy from the University of Arizona, said the signal was "99.9 percent likely to be human radio interference". For example, some electronics use 982.000 MHz oscillators, so perhaps the signal came from an oscillator that had drifted by a few parts per million, although why it should come from a particular direction in the sky remains a mystery.

If it does prove to be an artificial signal from Proxima Centauri, then it would be an astonishing coincidence that the first such indication of aliens should be from the star nearest to us!

As it turns out, Proxima Centauri was already of interest for another project under the auspices of Yuri Milner's Breakthrough Initiatives. In 2016 he funded studies for an effort to use a powerful laser to accelerate a large number of ultra-lightweight, centimetre-sized probes equipped with 'light sails' to nearby stars at 20% the speed of light. Such a probe would be able to reach Proxima Centauri in a couple of decades.

Further Reading
The Guardian report
www.theguardian.com/science/2020/dec/18/scientists-looking-for-aliens-investigate-radio-beam-from-nearby-star

Breakthrough Listen
en.wikipedia.org/wiki/Breakthrough_Listen

SETI Institute
www.seti.org

Proxima Centauri
en.wikipedia.org/wiki/Proxima_Centauri

December

Last Quarter: 5 December
New Moon: 12 December
First Quarter: 19 December
Full Moon: 27 December

MERCURY is visible in the west for most of the month, losing altitude steadily until it disappears just before inferior conjunction on 22 December. It reappears as a morning sky object as the year draws to a close. Mercury passes through its ascending node on 15 December and reaches perihelion five days later. It encounters the red planet for the second time this year when the two bodies are less than 4° apart on 28 December.

VENUS is on the decline in the east before sunrise as seen from northern latitudes but continues to climb above the horizon when viewed from southern summer skies. When observed through a telescope, Venus appears as a gibbous disk which gets smaller with every passing day, and over the course of the month, the morning star dims slightly from magnitude −4.2 to −4.1.

EARTH is at solstice on 22 December, welcoming summer to the southern hemisphere and bringing winter to the north. The Geminids, arguably the best meteor shower of the year, peak mid-month around the time of New Moon. The Ursids reach maximum activity about ten days later. For more information, see *Meteor Showers in 2023*. The Moon occults Neptune one final time this year. The occultation of Antares (α Scorpii) on 12 December is unobservable but the Moon's passage through the Pleiades (M45) on 24 December is easily visible.

MARS moves into the morning sky but is very close to the eastern horizon when the Sun appears and may not be observable this month. It begins December in Scorpius before moving into the non-zodiacal constellation of Ophiuchus on the fifth. It returns to the zodiac on the last day of the year when it enters Sagittarius. The red planet is beginning to brighten post-conjunction and outshines nearby Mercury on 28 December.

4 VESTA arrives at opposition with respect to both ecliptic longitude and right ascension on 21 December, with its greatest elongation from the Sun (177°) occurring the following day. This minor planet is found in the constellation of Orion but optical aids will be required to see it as the asteroid is only seventh magnitude. The article *Minor Planets in 2023* has more details, including a finder chart for the asteroid.

37 FIDES also comes to opposition this month. Smaller and fainter than 4 Vesta, it reaches opposition in ecliptic longitude on 18 December, a date which coincides with its opposition in right ascension and its greatest elongation from the Sun (175°). This tenth-magnitude asteroid is located in Auriga. For more information on this tiny solar system body, see *Minor Planets in 2023*.

JUPITER is an evening sky object, not setting until well after midnight. Found in Aries, the giant planet is 2.6° south of the waxing gibbous Moon on 22 December before moving from retrograde to direct motion on the last day of the year.

SATURN sets mid- to late evening this month. The planet shines at magnitude +0.8 in Aquarius which favours southern hemisphere planet watchers. The waxing crescent Moon glides past on 17 December, just 2.5° south of the planet.

URANUS ends the year as it began, an evening sky object in retrograde through the constellation of Aries. Now past opposition, Uranus is already above the horizon once the skies grow dark, and sets during or before morning twilight. The waxing gibbous Moon makes its final 2023 pass by the planet on 23 December when the two bodies are 2.7° apart.

NEPTUNE returns to direct motion early in the month, and leaves Aquarius for Pisces on 11 December. East quadrature takes place on 17 December. A third lunar occultation occurs two days later; like the two September occultations, this is visible primarily in Antarctica. Neptune is already above the horizon by the time darkness falls and sets around midnight.

A Funny Thing Happened on the Way to Seattle
Stargazing at 36,000 feet

David Harper

The cabin of a passenger plane may not seem to be an ideal place to observe the stars and planets. It is true that special flights are sometimes organised to watch total solar eclipses, and that NASA has equipped a Boeing 747 as an airborne infra-red observatory, taking advantage of the fact that airliners fly high above the layers of the Earth's atmosphere which block out much of the infra-red region of the electromagnetic spectrum. But most passengers on long-haul flights are more likely to while away the hours sleeping or watching in-flight movies than looking out of the window. That is a pity, because the view is often quite spectacular – mountains and lakes seen from seven miles high, or towering thunderclouds lit from within, or even an auroral display.

SOFIA – the Stratospheric Observatory for Infra-red Astronomy – is a modified Boeing 747 which carries a 2.7-metre reflecting telescope. The aircraft is seen here over the Sierra Nevada mountains with the telescope door open. (NASA/Jim Ross)

And sometimes, a traveller might witness an unexpected astronomical phenomenon which seems to defy explanation, such as the Full Moon rising in the north-north-east, then hovering a few degrees above the horizon and drifting slowly towards due north for almost five hours before setting.

I witnessed this curious event in December 2019 with my wife and fellow Yearbook contributor Lynne Marie Stockman, on a flight from London to Seattle. We had left Heathrow in the early afternoon on a ten-hour flight which took us across Iceland, southern Greenland and Baffin Island, passing far to the north of Hudson Bay and crossing the wilderness region of Canada called Nunavut. During the middle third of the flight, we were inside the Arctic Circle, and at our most northerly, we touched 70° N.

The very high latitude of our flight path is part of the explanation for the bizarre behaviour of the Moon. Within the Arctic Circle, the daily rising and setting of the Sun and Moon that is so familiar at lower latitudes can sometimes be interrupted. In high summer, the Sun may remain above the horizon for 24 hours a day, a phenomenon known as the Midnight Sun. In mid-winter, it can stay below the horizon for days or weeks or months. The Moon can also become circumpolar whenever it is far enough north of the celestial equator, such as the mid-winter Full Moon. At such times, it can appear above the horizon due north.

But even inside the Arctic Circle, the Sun, Moon and stars move from east to west when they are visible. The second clue to this astronomical puzzle lies in the fact that whilst we were inside the Arctic Circle, we were flying almost due west at 900 km/h (560 mph). At the equator, the meridian which marks local midday moves westward at 1,670 km/h (1,040 mph), so an aircraft flying along the equator would need to fly at almost one and a half times the speed of sound to keep up with the Sun. Concorde was able to do this on flights from Europe to New York: the flight time was as little as three hours, so travellers arrived in New York two hours "earlier" than when they left.

At higher latitudes, the meridians converge, allowing even non-supersonic aircraft to out-pace the Sun. At the Arctic Circle, an aircraft only needs to fly west at 660 km/h (410 mph) to follow the Sun, whilst at 70° N, a ground speed of 570 km/h (360 mph) is sufficient. We were flying significantly faster than this, so we were "overtaking" the Sun. If the aircraft had been equipped with a clock showing local true solar time, it would have been running backwards whilst we were north of 60° N. A clock showing local sidereal time would also have run backwards, and this is why the Moon moved from right to left along the northern horizon. We were flying west faster than the rotation of the Earth itself.

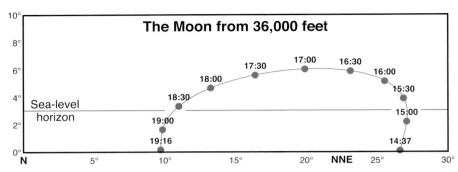

This diagram shows the Moon's path as seen by the author on 12 December 2019 on a flight from London to Seattle. Times are GMT. The Moon's disk is shown to scale. (David Harper)

Our altitude, 36,000 feet above sea level, was a third factor which helped to prolong the unusual phenomenon. At this altitude, a geometric effect known as "horizontal dip" allows the observer to see celestial objects that are below the horizon for an observer at sea level. The effect increases with altitude, and at 36,000 feet, it amounts to 3°. This brought the Moon into view half an hour before it would have been visible if we had been flying at sea level, and kept it above our local horizon for an extra half hour as it began set.

Comets in 2023

Neil Norman

Comets are generally regarded as being the most unpredictable objects of the Solar System and may appear at any given moment with no prior warning, as was shown by the 'great comet' of 2020, C/2020 F3 (NEOWISE). Discovered in March of that year, it went on to become a naked-eye object in July with several degrees of tail visible in a twilight sky. The next great comet could in fact be discovered at any time, and it is advisable to keep a close eye on social media or the British Astronomical Association (BAA) Comet Section page **www.ast.cam.ac.uk/~jds** where the latest comet discoveries are announced.

Best Prospects For 2023

At the time of writing, a total of 65 comets are due to return to perihelion during 2023, with 53 of these belonging to the Jupiter family of comets, four being long-period and the remaining eight being defunct comets or objects not observed for several returns. These missing objects are of interest because given the current technology available to amateur astronomers anyone could possibly recover one of these long-lost comets and perhaps gain some notoriety for themselves.

For those of you interested in hunting these objects, the following table gives information on the missing comets, the data shown relating to when they are expected to be at perihelion.

ERIHELION	COMET	RA	DECLINATION	MAGNITUDE	CONSTELLATION
Apr 2023	D/1978 R1	01 10 41	+05 01 50	15.5	Pisces
Apr 2023	5D/Brorsen	03 24 21	+22 58 18	5.6	Aries
Aug 2023	322P/SOHO	09 48 28	+13 41 01	6	Leo
Sep 2023	79P/du Toit–Hartley	12 51 00	−07 15 04	17.3	Virgo
Oct 2023	321P/SOHO	13 51 48	−08 44 27	?	Virgo
Nov 2023	P/2007 T2	18 16 59	−25 21 15	17.6	Sagittarius
Dec 2023	3D/Biela	20 28 07	−13 01 51	17.2	Capricornus
Dec 2023	147P/Kushida–Muramatsu	05 29 29	+22 35 13	17.7	Taurus

321P/SOHO is believed to be a dead comet having expelled all of its volatile materials and thus a magnitude cannot be given with any confidence.

Brightest Prospects for 2023

Although only six comets are expected to attain magnitude 10 or brighter, these objects are famed for their unpredictability, and outbursts could occur leading to one or more of them becoming brighter at any time. In any case, the following comets should be within the reach of the average astronomer's equipment.

237P/LINEAR

This comet was found to be a comet when initially discovered by the Wide-field Infrared Survey Explorer (WISE) on 10 June 2010. After checking it was found to have been previously recorded as an asteroidal object, discovered by the Lincoln Near-Earth Asteroid Research (LINEAR) project on 6 June 2002. The comet has an orbital period of 7.2 years and a perihelion distance of 1.98 au. At the return of 2016, the comet brightened quite rapidly and, although only expected to attain magnitude 13, it eventually reached a magnitude of 10 with a gradual dimming post-perihelion. In 2023 the comet is predicted to reach magnitude 11 at best, but could easily sneak into single digit numbers. This is one to watch very closely.

DATE	RA	DECLINATION	MAGNITUDE	CONSTELLATION
1 Jun 2023	19 59 50	−07 48 28	11	Aquila
15 Jun 2023	20 00 27	−04 37 10	11	Aquila
1 Jul 2023	19 54 11	−01 36 50	11	Aquila
15 Jul 2023	19 44 59	−00 06 43	11	Aquila

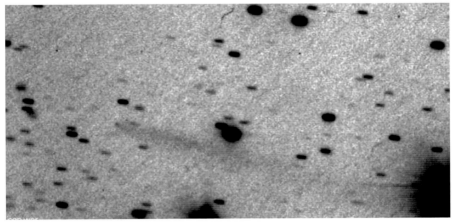

This image of Comet 237P/LINEAR was taken on 30 August 2018 from San Pedro de Atacama Observatory, Chile. The comet appears as a faint trail due to its lack of dust, a result of repeated perihelion passes around the Sun. (Alain Maury/Jean Gabriel Bosch)

103P/Hartley

Discovered in 1986 by the English astronomer Malcolm Hartley at the Schmidt Telescope Unit, Siding Spring Observatory, Australia, Comet 103P/Hartley could easily have an entire chapter devoted to it. This comet was studied up close when the Deep Impact spacecraft visited it on 4 November 2010, with a flyby distance of just under 700 kilometres.

In August 2008, the Spitzer Space Telescope showed the comet to have a nucleus approximately 1.6 kilometres in diameter. It has an albedo of 0.028, resulting in 103P/Hartley being a very dark object. The mass of the comet is 300 megatonnes. Despite its small size (by cometary standards) the comet should survive another 100 perihelion passages around the Sun given its current mass loss during each passage.

Rolando Ligustri captured this image of 103P/Hartley and the Cone Nebula in Monoceros from New Mexico on 1 November 2010. (Rolando Ligustri)

DATE	RA	DECLINATION	MAGNITUDE	CONSTELLATION
15 Aug 2023	01 03 24	+36 12 39	9.7	Andromeda
1 Sep 2023	02 43 29	+42 31 01	8.4	Perseus
15 Sep 2023	04 35 35	+41 33 23	7.6	Perseus
1 Oct 2023	06 31 10	+31 09 23	7.1	Auriga
15 Oct 2023	07 38 39	+18 55 26	7.1	Gemini
1 Nov 2023	08 28 43	+05 56 57	7.7	Hydra
15 Nov 2023	08 52 39	−02 13 53	8.4	Hydra

C/2020 V2 (ZTF)

This comet follows a hyperbolic trajectory meaning that it is making its first and only perihelion passage through the solar system before it is eventually kicked out of our system forever. At the peak of the apparition the comet could reach tenth magnitude when it arrives at perihelion in May this year at a distance of 2.2 au (beyond the orbit of Mars) from the Sun. The uncertainty lies in the fact that the comet may have been in outburst when discovered as a nineteenth magnitude object on 2 November 2020, the discovery being made by the Zwicky Transient Facility (ZTF) survey, Palomar Observatory, California, U.S.A.

DATE	RA	DECLINATION	MAGNITUDE	CONSTELLATION
15 Jun 2023	03 03 04	+18 28 40	10	Aries
1 Jul 2023	03 09 34	+15 00 13	10	Aries
15 Jul 2023	03 12 26	+11 13 19	10	Aries
1 Aug 2023	03 10 43	+05 12 20	10	Cetus
15 Aug 2023	03 03 15	−01 21 47	10	Eridanus

2P/Encke

This year sees the return of 'Old Faithful', Encke's Comet, first seen by French astronomer Pierre Méchain in 1786 but not recognised as a periodic comet until 1819 when its orbit was computed by German astronomer Johann Franz Encke. 2P/Encke has an orbital period of just 3.3 years, which means that it can be followed throughout the entire length of its orbit. Its brightness has reduced since it was first discovered, dropping from magnitude 3.5 to a current value of around 6 at best.

The parent of the Taurid and Beta Taurid meteor showers seen from here on Earth, 2P/Encke is also believed to be responsible for a meteor shower that occurs on Mercury. In addition, the Near-Earth object 2004 TG_{10} (responsible for the Northern Taurids) may be a fragment of Encke's Comet. The 2023 apparition favours the northern hemisphere with the comet being visible in the morning sky although elongation from the Sun from mid-October makes 2P/Encke a difficult object to visually observe during that period.

144P/Kushida imaged by Peter Carson on 5 November 2016 by Peter Carson from Leigh on Sea, Essex, England. (Peter Carson)

DATE	RA	DECLINATION	MAGNITUDE	CONSTELLATION
15 Sep 2023	08 01 08	+31 29 27	10.8	Gemini
1 Oct 2023	10 18 42	+18 46 38	8.3	Leo
15 Oct 2023	12 05 08	+02 11 28	5.7	Virgo

144P/Kushida

The Japanese amateur astronomer Yoshio Kushida discovered this object in a photograph taken in early January 1994 when the comet was at magnitude 13.5. The return of 1994 was a favourable one with a close approach to Earth on 13 January of 0.49 au. Belonging to the Jupiter Family group of comets, 144P/Kushida has an orbital period of just 7.45 years. This year the comet will approach the Earth to within 0.57 au in December.

DATE	RA	DECLINATION	MAGNITUDE	CONSTELLATION
1 Dec 2023	02 48 14	+16 50 49	9.9	Aries
15 Dec 2023	02 49 56	+15 15 15	9.6	Aries
1 Jan 2024	03 05 17	+14 23 52	9.4	Aries

12P/Pons-Brooks

Although this comet returns to perihelion in April of 2024, it is regarded as an exciting prospect for many astronomers and is ranked second only to Halley's Comet for objects of its type. With a period of 70.85 years, 12P/Pons-Brooks belongs to the Halley-type family of comets and was last at perihelion in May 1954.

First discovered on 21 July 1812 as a magnitude 6.5 object near the Camelopardalis-Lynx border by French astronomer Jean-Louis Pons, it attained naked-eye visibility

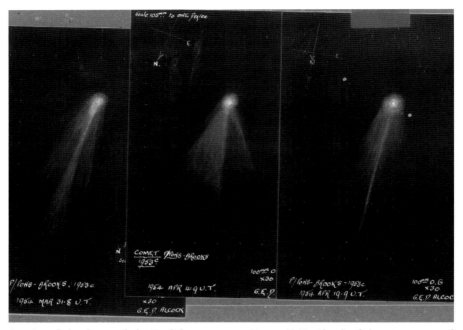

A series of sketches made by English astronomer George E. D. Alcock of the 1954 return of Comet 12P/Pons-Brooks. (George E. D. Alcock)

by mid-August, reaching fourth magnitude in mid-September, at which time it had a split tail with each tail section spanning around three degrees of sky.

The next return was that of 1883, when William Robert Brooks accidentally recovered it on 2 September during a general search for comets. Perihelion occurred in January 1884, when the comet reached third magnitude and sporting a tail 20 degrees long.

American astronomer Elizabeth Roemer next recovered the comet on 20 June 1953 on its way to a perihelion passage on 22 May 1954. This return was less favourable and a peak brightness of just magnitude 6 was recorded. However, an interesting fact about this comet is that it undergoes outbursts, during which the features of the comet can change rapidly. This has been seen on each of its returns, so anything is possible. The forthcoming return is much like that of 1954, but given the technology currently available to amateur astronomers this could be a very interesting event.

DATE	RA	DECLINATION	MAGNITUDE	CONSTELLATION
1 Dec 2023	18 26 07	+38 43 01	12	Lyra
15 Dec 2023	18 54 35	+38 00 05	11	Lyra
1 Jan 2024	19 36 46	+37 45 07	10	Cygnus

Minor Planets in 2023

Neil Norman

Minor planets are a collection of varying sized pieces of rock left over from the formation of the Solar system around 4.6 billion years ago. Millions of them exist, and to date almost 800,000 have been seen and documented, with around 550,000 having received permanent designations after being observed on two or more occasions and their orbits being known with a high degree of certainty. Different family types of asteroids also exist, such as Amor asteroids which are defined by having orbital periods of over one year and orbital paths that do not cross that of the Earth. Apollo asteroids have their perihelion distances within that of the Earth and thus can approach us to within a close distance and Trojan asteroids have their home at Lagrange points both 60 degrees ahead and behind the planet Jupiter respectively. These asteroids pose no problems to the Earth.

Most asteroids travel around the Sun in the main asteroid belt, which is located between the orbits of Mars and Jupiter. However, some asteroids have more elliptical orbits which allow them to interact with major planets, including the

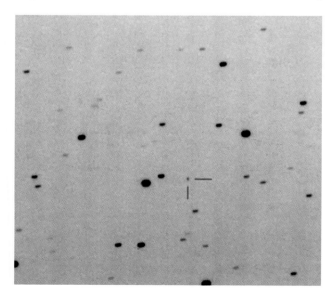

Potentially Hazardous Asteroid 99942 Apophis was crossing the constellation of Crater when it was imaged from Northolt Branch Observatories, London, England on 11 February 2021. (Guy Wells)

Earth. In all, there are around 2,000 which can approach to within a close distance of our planet. These objects are referred to as Potentially Hazardous Asteroids (PHAs). To qualify as a PHA these objects must have the capability to pass within 8 million km (5 million miles) of Earth and to be over 100 metres across.

Objects of this size could pose a serious threat if on a collision course with Earth. It is estimated that several thousand exist with diameters of over 100 metres, and with around 150 of these being over a kilometre across. A large number of smaller asteroids, measuring anything between just a few meters in diameter to several tens of meters wide, pass close to our planet on a regular basis, with considerable numbers of smaller ones entering the Earth's atmosphere every day, burning up harmlessly as meteors.

Those observers with a particular interest in following these objects should go to the home page of the Minor Planet Center. It is their job to keep track of these objects and determine orbits for them. This page can be accessed by going to **www.minorplanetcenter.net** where you will find a table of newly discovered minor planets and Near Earth Objects (NEOs). At the top of the page is a search box that you can use to locate information on any object that you are interested in, and from this you can obtain ephemeredes of the chosen subject. The Minor Planet Center site is the one that all dedicated asteroid observers should consult on a regular basis.

This year we are featuring asteroids that are at their greatest elongations from the Sun during 2023 and thus well placed for observation.

1 Ceres

This is no doubt the star of the show and arguably the best known of the minor planets. With a diameter of 945 km (587 miles) Ceres is the largest object in the asteroid belt. Discovered by Italian astronomer Giuseppe Piazzi from Palermo Observatory, Sicily on 1 January 1801, Ceres was initially believed to be a planet until the 1850s when it was reclassified as an asteroid following the discoveries of many other objects in similar orbits.

Ceres is at greatest elongation from the Sun on 21 March in the constellation of Coma Berenices.

DATE	R.A.	DEC	MAG	CONSTELLATION
1 Mar 2023	12 42 41	+13 22 58	7.2	Coma Berenices
15 Mar 2023	12 33 10	+14 52 27	7.0	Coma Berenices
21 Mar 2023	12 28 12	+15 24 51	6.9	Coma Berenices
1 Apr 2023	12 18 42	+16 07 20	7.0	Coma Berenices

2 Pallas

Pallas has a diameter of 512 km (318 miles) and was the second asteroid to be discovered when first spotted by the German astronomer Heinrich Wilhelm Matthias Olbers on 28 March 1802. He named the object after the Greek goddess of wisdom and warfare Pallas Athena, an alternative name for the goddess Athena. Pallas travels around the Sun once every 1,686 days, its orbit being highly eccentric, and its path around the Sun steeply inclined to the main plane of the asteroid belt, rendering it fairly inaccessible to spacecraft.

Pallas is at greatest elongation from the Sun on 16 January 2023 in the constellation of Canis Major.

DATE	R.A.	DEC	MAG	CONSTELLATION
1 Jan 2023	06 55 33	−32 00 52	7.7	Canis Major
15 Jan 2023	06 43 44	−30 17 28	7.7	Canis Major
16 Jan 2023	06 42 56	−30 05 50	7.7	Canis Major
1 Feb 2023	06 33 20	−25 52 09	7.7	Canis Major
15 Feb 2023	06 31 04	−20 53 10	7.8	Canis Major

4 Vesta

Discovered by German astronomer Heinrich Wilhelm Matthias Olbers on 29 March 1807, Vesta is one of the largest of the asteroids, with a diameter of 525 kilometres (326 miles) and an orbital period of 3.63 years. Vesta holds the distinction of being the brightest minor planet visible from Earth. With a maximum magnitude of 6, Vesta is the only one of the minor planets which is regularly bright enough to be seen with the naked eye.

Vesta is at greatest elongation from the Sun on 22 December 2023 in the constellation of Orion.

DATE	R.A.	DEC	MAG	CONSTELLATION
1 Dec 2023	06 18 18	+19 42 22	6.6	Orion
15 Dec 2023	06 04 35	+20 15 13	6.3	Orion
22 Dec 2023	05 56 46	+20 33 14	6.1	Orion

15 Eunomia

Discovered on 29 July 1851 by the Italian astronomer Annibale de Gasparis, and named after the Greek goddess of law and legislation, Eunomia has an orbital period of 4.3 years and appears to be a regularly shaped but elongated object with a mean diameter of around 232 kilometres (144 miles). It is the largest of the stony asteroids, containing approximately 1% of the mass of the asteroid belt.

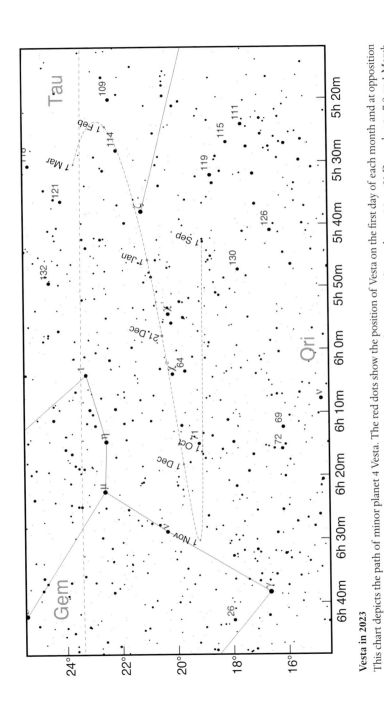

Vesta in 2023

This chart depicts the path of minor planet 4 Vesta. The red dots show the position of Vesta on the first day of each month and at opposition on 21 December. They are scaled to indicate the magnitude, which varies from 7.1 on 1 September to 6.1 on 21 December to 7.8 on 1 March 2024. Stars are shown to magnitude 9.0, which is the limit for binoculars or a small telescope. (David Harper)

Eunomia is at greatest elongation from the Sun on 8 July 2023 in the constellation of Sagittarius.

DATE	R.A.	DEC	MAG	CONSTELLATION
1 Jul 2023	19 14 09	−25 25 04	8.8	Sagittarius
8 Jul 2023	19 06 45	−25 08 59	8.6	Sagittarius
15 Jul 2023	18 59 15	−24 50 02	8.7	Sagittarius

37 Fides

The S-type (stony) main belt asteroid 37 Fides has a diameter of around 110 kilometres (68 miles) and was named after the Roman goddess of loyalty. It was discovered on 5 October 1855 by the German astronomer Karl Theodor Robert Luther from the Düsseldorf-Bilk Observatory in Germany, and was one of 24 asteroid discoveries made by Luther from this site. These objects are collectively known as the *24 Düsseldorf planets*.

Fides is at greatest elongation from the Sun on 18 December 2023 in the constellation of Auriga.

DATE	R.A.	DEC	MAG	CONSTELLATION
1 Dec 2023	05 59 37	+28 37 00	10	Auriga
15 Dec 2023	05 46 06	+28 50 03	9.7	Auriga
18 Dec 2023	05 42 55	+28 50 08	9.6	Auriga

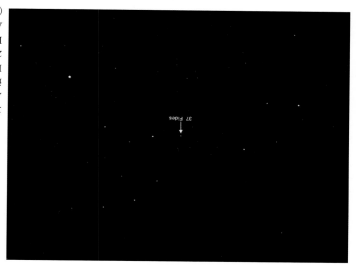

37 Fides was in Aquarius when imaged by Loren Ball on 5 October 2018 from Emerald Lane Observatory, Alabama, USA. (Loren Ball).

Meteor Showers in 2023

Neil Norman

A shooting star dashing across the sky is a wonderful sight that often captures the imagination of many young minds and perhaps sparks an interest for astronomy in them. On any given night of the year you can expect to see several of these, and they belong to two groups – sporadic and shower. Quite often the ones you see will be 'sporadic' meteors, that is to say they can appear from any direction and at any time during the observing session. These meteors arise when a meteoroid – perhaps a particle from an asteroid or a piece of cometary debris orbiting the Sun – enters the Earth's atmosphere and burns up harmlessly high above our heads, leaving behind the streak of light we often refer to as a "shooting star". The meteoroids in question are usually nothing more than pieces of space debris that the Earth encounters as it travels along its orbit, and range in size from a few millimetres to a couple of centimetres in size. Meteoroids that are large enough to at least partially

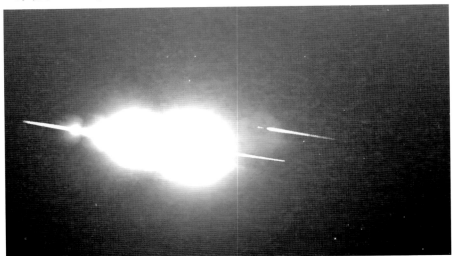

This image of a large fireball was captured on 28 February 2021 from Hullavington, Wiltshire, England. The object was seen to explode, and is believed to have left fragments scattered over a wide area of Gloucestershire. (Paul Dickinson/Global Meteor Network)

survive the passage through the atmosphere, and reach the Earth's surface without disintegrating, are known as meteorites.

At certain times of the year the Earth encounters more organised streams of debris that produce meteors over a regular time span and which seem to emerge from the same point in the sky. These are known as meteor showers. These streams of debris follow the orbital paths of comets, and are the scattered remnants of comets that have made repeated passes through the inner solar system. The ascending and descending nodes of their orbits lie at or near the plane of the Earth's orbit around the Sun, the result of which is that at certain times of the year the Earth encounters and passes through a number of these swarms of particles.

The term 'shower' must not be taken too literally. Generally speaking, even the strongest annual showers will only produce one or two meteors a minute at best, this depending on what time of the evening or morning that you are observing. One must also take into account the lunar phase at the time, which may significantly influence the number of meteors that you see. For example, a full moon will probably wash out all but the brightest meteors.

The following is a table of the principle meteor showers of 2023 and includes the name of the shower; the period over which the shower is active; the Zenith Hourly Rate (ZHR); the parent object from which the meteors originate; the date of peak shower activity; and the constellation in which the radiant of the shower is located.

Most of the information given is self-explanatory, but the Zenith Hourly Rate may need some elaborating.

The Zenith Hourly Rate is the number of meteors you may expect to see if the radiant (the point in the sky from where the meteors appear to emerge) is at the zenith (or overhead point) and if observing conditions were perfect and included dark, clear and moonless skies with no form of light pollution whatsoever. However, the ZHR should not be taken as gospel, and you should not expect to actually observe the quantities stated, although 'outbursts' can occur with significant activity being seen.

The observer can make notes on the various colours of the meteors seen. This will give you an indication of their composition; for example, red is nitrogen/oxygen, yellow is iron, orange is sodium, purple is calcium and turquoise is magnesium. Also, to avoid confusion with sporadic meteors which are not related to the shower, trace the path back of the meteor and if it aligns with the radiant you can be sure you have seen a genuine member of the particular shower.

Meteor Showers in 2023

SHOWER	DATE	ZHR	PARENT	PEAK	CONSTELLATION
Quadrantids	1 Jan to 5 Jan	120	2003 EH$_1$ (asteroid)	3/4 Jan	Boötes
Lyrids	16 Apr to 25 Apr	18	C/1861 G1 Thatcher	22/23 Apr	Lyra
Eta Aquariids	19 Apr to 28 May	30	1P/Halley	6/7 May	Aquarius
Tau Herculids	19 May to 19 Jun	Varies	73P/Schwassmann–Wachmann	9 Jun	Hercules
Delta Aquariids	12 Jul to 23 Aug	20	96P/Machholz	28/29 Jul	Aquarius
Perseids	17 Jul to 24 Aug	80	109P/Swift-Tuttle	12/13 Aug	Perseus
Draconids	6 Oct to 10 Oct	10	21P Giacobini-Zinner	8/9 Oct	Draco
Southern Taurids	10 Sep to 20 Nov	5	2P/Encke	10 Oct	Taurus
Orionids	2 Oct to 7 Nov	20	1P/Halley	21/22 Oct	Orion
Northern Taurids	20 Oct to 10 Dec	5	2004 TG$_{10}$ (asteroid)	12 Nov	Taurus
Leonids	6 Nov to 30 Nov	Varies	55P/Tempel/Tuttle	17/18 Nov	Leo
Geminids	4 Dec to 17 Dec	75+	3200 Phaethon (asteroid)	13/14 Dec	Gemini
Ursids	17 Dec to 26 Dec	10	8P/Tuttle	22/23 Dec	Ursa Minor

Quadrantids

The parent object of the Quadrantids has been identified as the near-Earth object of the Amor group of asteroids 2003 EH$_1$, which is likely to be an extinct comet. The radiant lies a little to the east of the star Alkaid in Ursa Major and the meteors

This Quadrantid meteor was captured on 4 January 2016 by Mary McIntyre from Oxfordshire, England. The constellation of Leo is visible along with the planet Jupiter. (Mary McIntyre)

are fast moving, reaching speeds of 40 km/s. Maximum occurs on the night of 3/4 January when most of the fainter meteors will be blocked out by a nearly full Moon.

Lyrids

Produced by particles emanating from the long-period comet C/1861 G1 Thatcher, these are very fast moving meteors that approach speeds of up to 50km/s. The peak falls on the night of 22/23 April with the radiant lying near the prominent star Vega in the constellation of Lyra. This year a thin crescent Moon will have set early in the evening and the skies will be dark enough to reveal a potentially excellent display.

Eta Aquariids

One of the two showers associated with 1P/Halley, the Eta Aquariids are active for a full month between 19 April and 28 May. The radiant lies just to the east of the star Sadalmelik (α Aquarii), from where up to 30 meteors per hour are normally expected during the period of peak activity, although displays of up to 60 meteors per hour can occasionally be seen. Maximum activity occurs in the pre-dawn skies of 7 May when a Moon just past full phase will obscure all but the brightest meteors from this shower.

Tau Herculids

Appearing to originate from the star Tau (τ) Herculis, this shower runs from 19 May to 19 June with peak activity taking place on 9 June. The Tau Herculids were first recorded in May 1930 by observers at the Kwasan Observatory in Kyoto, Japan. The parent body has been identified as the comet 73P/Schwassmann–Wachmann, discovered on 2 May 1930 by the German astronomers Arnold Schwassmann and Arno Arthur Wachmann during a photographic search for minor planets being carried out from Hamburg Observatory in Germany.

73P/Schwassmann–Wachmann has an orbital period of 5.36 years. During 1995 the comet began to fragment, and by the time of its 2006 return, at least eight individual fragments were observed (although the Hubble Space Telescope spotted dozens more). 73P/Schwassmann–Wachmann appears to be close to total disintegration. The observed rate of meteors from this shower are low, but with this said, recent research suggests that Earth will be getting closer to the streams of debris left behind by the fragmenting comet and thus heavier meteor showers can be expected in the coming years with another hefty shower possible for 2049. Peak activity in 2023 occurring just before the last quarter Moon should ensure a reasonably good display.

Delta Aquariids

Probably linked to the short-period sungrazing comet 96P/Machholz, the Delta Aquariids is a fairly average shower which coincides with the more prominent Perseids. However, Delta Aquariid meteors are generally much dimmer than those associated with the Perseids, making their identification somewhat easier. The radiant lies to the south of the Square of Pegasus, close to the star Skat (δ Aquarii). Located to the north of the bright star Fomalhaut in Pisces Austrinus, the radiant is particularly well placed for those observers situated in the southern hemisphere. The shower peaks during the early hours of 29 July, when the Moon is approaching full phase, so most of the fainter meteors may be obscured.

Perseids

Associated with the parent comet 109P/Swift-Tuttle, the Perseids are a beautiful sight, with meteors appearing as soon as night falls and up to 80 meteors per hour often recorded. This is usually one of the best meteor showers to observe, large numbers of very bright meteors often being seen, and at the time of peak activity on the night/morning of 12/13 August, a waning crescent Moon should leave skies dark enough to allow for a reasonably good show.

A bright Perseid meteor streaks across the skies above Oxfordshire, England on 12 August 2017. (Mary McIntyre)

Southern Taurids/Northern Taurids

Running collectively from 10 September to 10 December, the Taurids are two separate showers, with southern and northern components, the Southern Taurids linked to the periodic comet 2P/Encke and the Northern Taurids to the asteroid 2004 TG_{10}. The southern hemisphere encounters the first part of the stream, followed later by the northern hemisphere encountering the second part. The ZHR of these showers is low (between 5 and 10 per hour), although they can be beautiful to watch as they glide across the sky. The Southern Taurids peak on 10 October when a waning crescent Moon will result in dark skies and the potential for a decent display. A New Moon on the night of 12 November should ensure that the Northern Taurids fare even better.

Draconids

Also known as the Giacobinids, the Draconid meteor shower emanates from the periodic comet 21P/Giacobini-Zinner. The duration of the shower is just 4 days from 6 October to 10 October, with the shower peaking on 8/9 October. The ZHR of this shower varies with poor displays in 1915 and 1926 but stronger displays in 1933, 1946, 1998, 2005 and 2012. Radiating from the "head" of Draco, the meteors from this shower travel at a relatively modest 20 km/s. The Draconids are generally quite faint, although a thin crescent Moon will result in dark skies and the potential for a good show.

Orionids

The second of the meteor showers associated with 1P/Halley, the Orionids radiate from a point a little to the north of the star Betelgeuse in Orion. Best viewed in the early hours when the constellation is well placed, the shower takes place between 2 October and 7 November with a peak on the night of 21/22 October. The velocity of the meteors entering the atmosphere is a speedy 67 km/s. This year, a first quarter Moon may drown out some of the fainter members of this shower, although the post-midnight skies will be free of moonlight and will leave dark skies and a rewarding display of meteors.

Leonids

Running from 6 November to 30 November, this is a fast moving shower with particles varying greatly in size and which can create lovely bright meteors that occasionally attain magnitude −1.5 (or about as bright as Sirius) or better. The radiant is located a few degrees to the north of the bright star Regulus in Leo. The parent of the Leonid shower is the periodic comet 55P/Tempel-Tuttle which orbits the Sun every 33 years. It was last at perihelion in 1998 and is due to return in May 2031. The Zenith Hourly Rates vary due to the Earth encountering material from different perihelion passages of the parent comet. For example, the storm of 1833 was due to the 1800 passage, the 1733 passage was responsible for the 1866 storm and the 1966 storm resulted from the 1899 passage (for additional information see Courtney Seligman's article *Cometary Comedy and Chaos* in the *Yearbook of Astronomy 2020*). The Leonid shower peaks on the night of 17/18 November while the Moon is approaching first quarter but will have set by midnight, ensuring dark skies and a decent early morning display.

Geminids

The Geminid meteor shower was originally recorded in 1862 and originates from the debris of the asteroid 3200 Phaethon. Discovered in October 1983, this rocky five kilometre wide Apollo asteroid made a relative close approach to Earth on 10 December 2017, when it came within 0.069 au (10.3 million kilometres/6.4 million miles) of our planet. The Geminid radiant lies near the bright star Castor in Gemini and the shower peaks on the night of 13/14 December. This is considered by many to be the best shower of the year, and it is interesting to note that the number of observed meteors appears to be increasing annually. This year a New Moon at the time of peak activity will result in dark skies and the potential for really good display.

Ursids

Discovered by William Frederick Denning during the early twentieth century, this shower is associated with comet 8P/Tuttle and has a radiant located near Beta (β) Ursae Minoris (Kochab). With relatively low speeds of 33 km/s, the Ursids are seen to move gracefully across the sky. The shower runs from 17 December to 26 December with peak activity taking place on the night of 22/23 December when many of the fainter meteors will be drowned out by the light from a waxing gibbous moon.

The English amateur astronomer William Frederick Denning is best remembered for his five comet discoveries and his catalogues of meteor shower radiants. (John McCue)

Andromedids

The Andromedids are now a very weak shower with meteors rarely even reaching naked eye visibility. However, perturbations of the stream's node means there is a possibility that the Andromedid meteor shower may explode into life in 2023. For further information on the shower and its origins, see the article *Biela's Comet: Life After Death?* elsewhere in this volume.

Article Section

Recent Advances in Astronomy

Rod Hine

More Insights from the Cosmic Microwave Background Radiation

Ever since its chance discovery in 1965 by Arno Penzias and Robert Wilson, the cosmic microwave background radiation (CMB) has been the subject of much research seeking clues to the very early development of the universe. Its very existence provides one of the main pillars to support the Big Bang theory as it consists of the drastically red-shifted radiation which was set free to fill the universe when the temperature fell to around 3,000K. At this moment, some 375,000 years after the big bang, electrons and protons combined to form hydrogen atoms and the universe became transparent.

In addition to being red shifted by a factor of about a thousand, the CMB carries evidence of interactions and events over the subsequent 13.8 billion years. One such effect arises from the light of the very early stars whose ultra-violet light would

The antenna and Low Noise amplifier for the EDGES experiment, at the Murchison Radio-astronomy Observatory in Western Australia. (Wikimedia Commons/Suzyj)

have penetrated the clouds of primordial hydrogen and excited the 21-centimetre hyperfine line. The excited gas would then absorb a small amount of energy from CMB photons at that wavelength. Of course, we make much use of the Hydrogen 1 line at a frequency of 1,420 MHz to map the current state of the universe, but since these early stars formed, the expansion of the universe has red shifted the CMB. So, in order to look for evidence of the absorption of the CMB we must examine it at much longer wavelength and lower frequency.

That work has been done by a team led by Judd D. Bowman of the School of Earth and Space Exploration, Arizona State University, Tempe, 85287, Arizona, USA. Given the name EDGES (Experiment to Detect the Global EoR Signature), the main instrument is a small ground-based radio antenna located in a radio-quiet region of Western Australia. However, working between 50 MHz and 100 MHz poses a daunting technical challenge. Even in the radio-quiet zone, the interference from both natural and man-made noise is of the order of a thousand times stronger than the tiny CMB signal. Only by incredibly precise calibration and the meticulous identification of all extraneous signals was it possible to make the observations.

The published results indicate a broad absorption profile centred on 78 MHz with a width of 19 MHz and amplitude of 0.5 Kelvin. With the red-shift z-factor of 18.2 it corresponds to barely 180 million years after the Big Bang. The results also indicated that the early universe was colder than expected at that age and this may be linked to possible interactions with dark matter.

Another new telescope project intended to work at similar extreme red-shifts is located in South Africa. Known as HERA (Hydrogen Epoch of Reionization Array), it consists of a hexagonal grid, 300 metres across, packed with no fewer than 350 14-metre diameter dishes aimed at the zenith in the arid Karoo Desert, not far from the MeerKAT telescope. Operating at 154 MHz – or about 2 metres wavelength – HERA's array will have sufficient sensitivity and resolution to image the large scale H1 structure of the early universe. In this respect it is quite different from EDGES which measured the sky-averaged spectrum.

Giant Radio Galaxies

Although many millions of radio galaxies are known, only a few hundred are classified as "giant radio galaxies" whose jets of radio waves extend across distances of the order of ten to one hundred times larger than the size of the Milky Way. These huge jets are, almost by definition, very faint and diffuse so have been difficult to detect. However, a large team of astronomers from around the world have been working with data collected by the MeerKAT radio telescope in South Africa in a collaboration called MIGHHTEE, otherwise known as the MeerKAT International

One of the array of 64 MeerKAT radio telescope antennas. (Wikimedia Commons / Morganoshell)

GHz Tiered Extragalactic Exploration. MeerKAT consists of 64 antennas in the Northern Cape district of South Africa. Each antenna is 13.5 metres in diameter and they are arranged within an eight kilometre diameter circle. Each antenna has cryogenic receivers and signals are immediately digitised and transmitted by optical fibre to a central processing building and thence to the operations centre in Cape Town.

Led by Jacinta Delhaize of the Department of Astronomy, University of Cape Town, South Africa, MIGHTEE reported the discovery of two truly gigantic radio galaxies, several billion light years distant and both about 6.5 million light years across – around 62 times the size of the Milky Way.

This could have implications for understanding how galaxies evolve over hundreds of millions of years, driven by super-massive black holes. Such huge radio jets and their associated huge outflows of ionised gas could influence further star formation by blowing away all the gas and inhibiting further star formation.

Fast Radio Bursts (FRBs) Detected at Low Frequencies

Since their initial discovery in 2007, FRBs have remained an enigma. Observed as transient radio pulses lasting less than a few milliseconds, they appear to originate

This image is an artist's impression representing the path of a typical fast radio burst travelling from a distant host galaxy to reach the Earth. Because of dispersion the higher frequencies travel very slightly faster and the initial short burst is spread out, arriving at the Earth as a "chirp" of descending frequency. (ESO / M. Kornmesser)

from some kind of extremely high-energy mechanism whereby the total energy in just one burst may correspond to as much energy as the Sun produces in several days. The energy is spread over a wide range of frequencies and a characteristic feature is the dispersion of the signal because the higher frequency components arrive slightly before the lower frequencies due to interactions with ionised plasma in the path of the signals.

Recent observations by researchers at McGill University, Montreal, Quebec, Canada, using the Canadian Hydrogen Intensity Mapping Experiment (CHIME) instrument, and the similar LOw-Frequency ARray (LOFAR) instrument in the Netherlands, has revealed that FRBs can be detected down to much lower frequencies than hitherto seen, with delays of up to three days between the higher frequencies detected by CHIME and the lower frequencies seen by LOFAR. This has redrawn the boundaries for those astrophysicists who are trying to understand the origins of FRBs as it says a lot about the environment surrounding the objects giving rise to the FRBs.

"We detected fast radio bursts down to 110 MHz where before these bursts were only known to exist down to 300 MHz," explained Ziggy Pleunis, a postdoctoral researcher in McGill's Department of Physics and lead author of the research published in *The Astrophysical Journal Letters*. "This tells us that the region around the source of the bursts must be transparent to low-frequency emission, whereas some theories suggested that all low-frequency emission would be absorbed right away and could never be detected." The search for the mechanism for producing FRBs goes on.

Closest Black Hole Yet Detected

Researchers at Ohio State University, led by Tharindu Jayasinghe, have discovered an unusually small black hole just 1,500 light years from Earth in the constellation of Monoceros (the Unicorn). With a mass just three times that of our Sun, it is companion to a red giant star which had been well documented in several surveys looking for planets around nearby stars. When the data was finally analysed, Tharindu and his group noticed that not only was the red giant wobbling as it orbited around something, its appearance and intensity changed at various points around the orbit. That doesn't normally happen in the case of planets and stars, but if the unseen companion were a black hole then the extreme tidal effects could cause the red giant star to assume a football-like shape with the longer axis lined up the black hole. That neatly explained the apparent changes in appearance. From the parent constellation, the black hole has been christened "The Unicorn" because it is, so far, just one of a kind. Spurred on by this discovery, the group at Ohio State are now searching in earnest for more small and medium-sized black holes within our region of the Milky Way.

Citizen Science

Advances in modern telescopes have produced huge amounts of data, far too much to be analysed by a few researchers, yet even the best computers and AI can't really compete with the human eye and brain for spotting objects. In particular, bodies known as "brown dwarfs" are of interest, are plentiful but as failed stars are notoriously dim

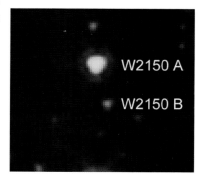

This image shows W2150AB, a wide (341 astronomical units projected distance) binary discovered by volunteers of the NASA project Backyard Worlds. This image was created with data from Spitzer Proposal ID #14076, using the 3.5 Micron image for cyan and the 4.5 Micron image for orange. (Wikimedia Commons/ NASA/JPL-Caltech Spitzer Space Telescope/Meli thev/IRSA/Spitzer Heritage Archive/Proposal 14076 + Own work/CC BY-SA 4.0)

in the infra-red and hard to spot. The NASA-funded project "Backyard Worlds; Planet 9" set up a collaboration between professional scientists and over 100,000 keen citizen scientists. The result has been published as a 3D map of brown dwarfs located within 65 light years of Earth.

"Without the citizen scientists, we couldn't have created such a complete sample in so short a time" said J. Davy Kirkpatrick, lead author of the study and scientist at the Infrared Processing and Analysis Center at the California Institute of Technology, in a NASA statement. "Having the power of thousands of inquiring eyes on the data enabled us to find brown dwarf candidates much faster." Once identified, the team was able to confirm the location of the objects and make further measurements using the Spitzer Space Telescope and other instruments.

The Galileo Project at Harvard

Launched with a public announcement on 26 July 2021, the Galileo Project intends to complement the more traditional SETI approach by using a multiplicity of modest telescopes to image unidentified aerial phenomena (UAP). The search will extend from objects within the Earth's atmosphere out to unusual objects such as 'Oumuamua which might enter the Solar System. The Head of the Project is

This artist's impression depicts 'Oumuamua, the first interstellar asteroid. Discovered on 19 October 2017 by the Pan-STARRS 1 telescope at Haleakalā Observatory in Hawaii, 'Oumuamua was the first known interstellar object to pass through the Solar System. (ESO/M Kornmesser/ nagualdesign)

Avi Loeb, currently Frank B. Baird Professor of Science and Director of the Harvard Astronomy Department, Cambridge, Massachusetts.

In an article in Scientific American on 19 September 2021, Avi Loeb explains the background for this unusual project. The topic of UAP has been dominated so far by very large numbers of anecdotal reports of UFOs from civilian witnesses, as well as the observations reported by the USA Office of the Director of National Intelligence (ODNI). Many of the ODNI observations remain classified for security reasons or perhaps just because they were recorded by government-owned sensors. The Galileo Project will make its full data set open to the scientific community and thus bring the whole potentially controversial field under rigorous scrutiny. "We aim to change the intellectual landscape of UAP studies by bringing them into the mainstream of credible scientific inquiry." says Loeb.

Whether or not Galileo will "solve the mystery of UFOs and interstellar objects", as suggested in at least one lurid news report, it does represent a serious attempt to analyse and understand the many strange phenomena that are not easily dismissed. The recent brief visit of 'Oumuamua – the first truly interstellar object known to visit the Solar System – posed many intriguing questions that are still not answered. Will they eventually discover "Extraterrestrial Artefacts"? The strap line on the Galileo Project home page says it all: "Daring to Look Through New Telescopes". One can only wish them success in what could be a very long and exhausting enterprise.

Recent Advances in Solar System Exploration

Peter Rea

This article was written in the summer of 2021. As the missions mentioned are either active or due for launch imminently, the status of some missions may change after the print deadline.

Double Asteroid Redirection Test (DART)

DART looks like being an intriguing mission. It will test the concept of using the kinetic energy of a colliding spacecraft to redirect an asteroid that could be on a collision course with Earth. The Applied Physics Laboratory of the Johns Hopkins University will manage the mission which will send DART to the binary asteroid 65803 Didymos. The larger of the two asteroids is approximately 780 metres in diameter and its smaller companion called Dimorphos is approximately 160 metres in diameter and orbits the primary at about 1 kilometre.

The DART spacecraft is essentially an impactor that will use its kinetic energy to impart a small but measurable change in velocity to the target. Whenever any spacecraft travels to the planets, tiny course corrections early in the mission can make a significant alteration to its trajectory. The same applies to the DART

Illustration showing DART on course to impact Didymos B (Dimorphos), viewed from behind the DART spacecraft. (NASA/Johns Hopkins APL)

mission. A small alteration along the velocity vector of the asteroid can make a big difference to its trajectory over time. The energy from the 500kg impactor could alter the orbital period of Dimorphos by a few minutes. If the asteroid were on a collision course with Earth and was impacted by a colliding spacecraft when it was still tens of millions of kilometres from Earth, that tiny change in velocity could be sufficient to make the asteroid pass by the Earth instead of colliding with it.

A secondary payload on this mission is being supplied by the Italian Space Agency. It is a small CubeSat carrying a camera to record the impact. Launch is scheduled for 24 November 2021 on a Falcon 9 rocket from Vandenberg Air Force Base in California. Impact with Dimorphos is expected in October 2022. The mission website is at **dart.jhuapl.edu**

Lucy in the Sky (with a few Trojan Asteroids)

When I first started in astronomy over 50 years ago the solar system seemed a simpler place somehow. We had nine planets, an asteroid belt and comets that seemed to come out of nowhere. Books on astronomy always showed the asteroid belt crowded with rocky bodies. This is a false picture; the asteroid belt is largely empty and many spacecraft have traversed this region without coming closer than a few million kilometres to an asteroid. Half a century on we realise the solar system is not just a simple asteroid belt lying between the orbits of Mars and Jupiter. We have categorised the asteroids into families based on their locations, orbital elements, and inclinations. One such family are the Trojans that sit at the L4 and L5 gravitationally stable Lagrange points at 60 degrees ahead and 60 degrees behind Jupiter's orbit. This family of asteroids have never been explored by

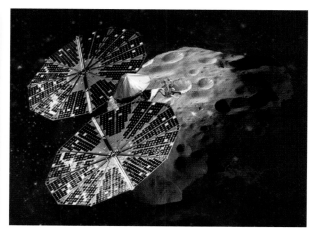

An artist's conception showing the Lucy spacecraft flying by the Trojan Eurybates – one of the six diverse and scientifically important Trojans to be studied. Trojans are fossils of planet formation and so will supply important clues to the earliest history of the solar system. (NASA / SwRI and SSL / Peter Rubin)

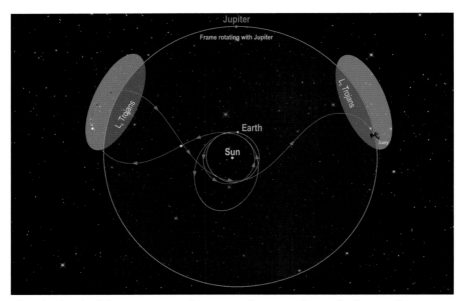

The orbital path of the Lucy spacecraft (green) is shown in a frame of reference where Jupiter remains stationary, giving the trajectory its pretzel-like shape. After launch in October 2021, Lucy has two close Earth flybys before encountering its Trojan targets. (NASA/Southwest Research Institute)

spacecraft. Lucy plans to change that. The name Lucy comes from the name given to a fossil find in the Afar Triangle Region of Hadar in Ethiopia by the American paleoanthropologist Donald Johanson of a female Australopithecus afarensis. The Trojan asteroids hopefully will provide "fossil" evidence of the early formation of the solar system. Let us hope Lucy the spacecraft provides as much information about our past as Lucy the skeleton did. Just over 3 million years separate the two!

Lucy is due for launch in October 2021 and will use two gravity assists of Earth to increase heliocentric velocity enabling Lucy to reach the orbit of Jupiter. On the way to the Trojans at the leading or L4 location the spacecraft will perform a flyby of the 4 kilometre diameter asteroid 52246 Donaldjohanson in 2025. Wait a minute. Back in 1974 Donald Johanson discovered Lucy and in 2025 Lucy will make discoveries at Donaldjohanson. Now that's comedy – or great planning!

In 2027 Lucy will arrive at the L4 location with the "Greek" camp of asteroids as most are named after Greek heroes. She will visit two asteroids that year followed by Eurybates and Polymele a month later. 2028 will see the flyby of Leucus and Orus, after which there will be a change of course. Lucy will return to the Earth for another gravity assist to send her onward to the L5 or trailing camp of Trojans in 2033 for a flyby of the binary asteroid Patroclus and Menoetius. This should be a fascinating mission to follow. The mission website can be found here: **lucy.swri.edu**

Psyche: Mission to a Metal World

Discovered by the Italian astronomer Annibale de Gasparis in 1852, asteroid Psyche lies in the main asteroid belt between the orbits of Mars and Jupiter, recent observations indicating it has an average diameter of 226 kilometres. Unlike most other asteroids, which are rocky or icy bodies, scientists believe the M-type (metallic) asteroid Psyche is comprised mostly of nickel and iron, rather like the Earth's core. A NASA mission of the same name is due to launch in August 2022. Using Mars for a gravity assist in May 2023 (See article "Gravity Assists – Something for

An artist's impression of the Psyche spacecraft, due for launch in 2022 toward a metallic asteroid of the same name. (NASA)

Nothing?" in the *Yearbook of Astronomy 2022*). The spacecraft should arrive at Psyche in late January 2026 and orbit the asteroid for almost two Earth years. The initial orbit would be about 700 kilometres above the surface of the asteroid, gradually being reduced to about 85 kilometres. This will be the first metallic asteroid ever to be explored. The mission can be followed using the website **psyche.asu.edu**

Rosalind Franklin goes to Mars

This section originally appeared in the *Yearbook of Astronomy 2021*. The Yearbook had gone to print prior to ESA postponing the mission until the 2022 launch opportunity.

As part of their ongoing ExoMars programme, the European Space Agency (ESA) in conjunction with Roscosmos in Russia will be sending a rover to Mars under the program name ExoMars 2022. The mission will comprise a stationary lander and a rover named 'Rosalind Franklin' after the English chemist and x-ray crystallographer whose work would assist in the discovery of the double helix structure of deoxyribonucleic acid (DNA). ESA has not yet successfully landed on Mars and readers may recall that back in 2016 the ESA Trace Gas Orbiter mission carried an experimental lander called Schiaparelli. The lander entered the Martian atmosphere on 16 October 2016. The initial entry went according to plan and a parachute was successfully deployed. However shortly after this, issues with the

The Rosalind Franklin rover on Mars showing the landing stage in the background. (ESA/ATG medialab)

computer caused an early release of the parachute and the retro propulsion system thrusters only fired for a few seconds before being commanded off. This caused the lander to impact the surface at 540 kilometres per hour. However much was learnt from this landing attempt and the reason for the failure is understood giving confidence for the 2022 lander.

The method of landing uses the same technique as the Schiaparelli lander, heat shield, parachute, and powered descent. Roscosmos will provide the lander which carries a significant suite of science instruments for in situ measurements. Data from these instruments will be relayed back to Earth via ESA's Trace Gas Orbiter. The rover, principally built in the UK at Airbus Defence and Space, is set for launch in September 2022. Landing is scheduled for June 2023 in the Oxia Planum region which lies east of Ares Vallis and Chryse Planitia where Viking 1 landed back in 1976. Once the rover is deployed onto the surface of Mars for independent studies the lander will pursue a stationary investigation of the landing site with its own suite of instruments. I think Rosalind Franklin would have been proud.

The mission website giving a comprehensive description of all the science instruments can be found at **exploration.esa.int/mars/48088-mission-overview**

Current Mars Missions Summary

Mars is often referred to as the Red Planet, and for very good reasons. When favourably placed it can dominate the sky and looks distinctly reddish in colour, caused by the dominance of iron oxide on the surface. Mars has certain similarities to Earth. It has pole caps, a similar axial tilt, and similar rotation period. But there the similarities end. Earth is warmer and three quarters covered in oceans with abundant free running water. Mars is a cold, desiccated world with a very thin atmosphere whose surface pressure is far too low for water to remain in liquid form. Observations by numerous spacecraft from orbit and from the surface have shown that Mars in the distant past was warmer, had a thicker atmosphere and had abundant water. For this and many other reasons, Mars had been a major destination for spacecraft over the past few decades. The accompanying table shows all active missions either in orbit or on the surface of Mars, and is correct as of the time of writing, summer 2021. It does not include the upcoming ESA led mission ExoMars 2022 described elsewhere in this article.

Mission	Origin	Type	Orbital Insertion	Landed
Mars Odyssey	USA	Orbiter	24 Oct 2001	
Mars Express	Europe	Orbiter	25 Dec 2003	
Mars Reconnaissance Orbiter	USA	Orbiter	10 Mar 2006	
Mars Science Laboratory – Curiosity	USA	Lander		6 Aug 2012
Maven	USA	Orbiter	22 Sep 2014	
Mars Orbiter Mission – Mangalyaan	India	Orbiter	24 Sep 2014	
ExoMars – Trace Gas Orbiter	Europe	Orbiter	19 Oct 2016	
InSight	USA	Lander		26 Nov 2018
Hope Mars Mission	UAE	Orbiter	9 Feb 2021	
Tianwen-1	China	Orbiter/ Lander	10 Feb 2021	14 May 2021
Perseverance Rover	USA	Lander		18 Feb 2021

The scope of these missions ranges from getting down among the rocks with two six-wheeled rovers (Curiosity and Perseverance) to a stationary lander listening for marsquakes (InSight). The rest are orbiters using a range of multispectral instruments for analysing the global surface composition (Mars Odyssey and Mars Express) and some carrying high resolution cameras for really closeup images (Mars Reconnaissance Orbiter). Others are dedicated to looking at the atmosphere of Mars (Maven and Trace Gas Orbiter). Arriving at Mars on 10 February 2021, the Chinese Tianwen-1 mission included an orbiter, lander and rover. On 14 May 2021, the lander successfully touched down on the surface of Mars in the Utopia Planitia region. The rover, named Zhurong, was successfully driven off the lander on 22 May 2021. Over the last 20 years – with the missions mentioned in the table – we have learned more about Mars than we have during the last 200 years of telescopic observations. They all pave the way to one day sending humans to Mars. Though some way off, when human footprints finally get planted on the red dusty surface of Mars, scientists can look back at the opening two decades of the twenty first century to the unmanned missions that made their landing possible.

Follow up information can be found at the following programme websites or by putting the mission's name into your search engine of choice. Just put the word "mission" after each name for a better return on searches.

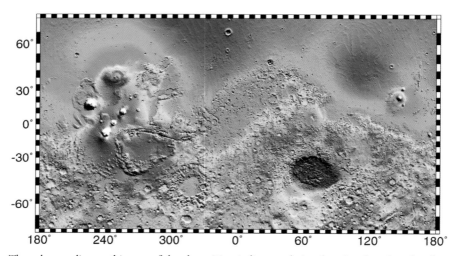

The colour coding on this map of the planet Mars indicates relative elevations based on data from the Mars Orbiter Laser Altimeter on NASA's Mars Global Surveyor. Red is higher elevation and blue is lower elevation. It clearly shows the north/south asymmetry. The northern hemisphere of Mars has a much lower elevation than the southern hemisphere. This view of Mars was impossible before spacecraft started to visit the planet. (NASA/JPL/GSFC)

ESA: **esa.int/Science_Exploration/Human_and_Robotic_Exploration/Exploration/ExoMars**

NASA: **mars.nasa.gov**

ESA to Jupiter with a Juicy Mission

Over the last few years it has become apparent that some of the large natural satellites of Jupiter and at least one natural satellite of Saturn – Enceladus – could well have liquid water below their surfaces. The joint NASA/ESA mission Cassini has explored in detail the Saturnian system. Two follow up missions to the icy moons of Jupiter are nearing launch. The NASA mission Europa Clipper is due for launch in 2024 and will be covered in later Yearbooks. In 2022 the European Space Agency will launch their Jupiter Icy Moons Explorer (JUICE) mission.

A major goal of the JUICE mission is to orbit Ganymede, something never achieved before. The long journey to Ganymede will start in the summer of 2022 with a launch from Kourou in French Guiana aboard an Ariane 5. With a launch mass of 4.8 tonnes even the powerful Ariane 5 cannot send JUICE directly to Jupiter. The spacecraft will need to gain additional energy from multiple planetary

JUICE is devoted to completing a unique tour of the Jupiter system and is the first large-class mission in ESA's Cosmic Vision 2015–2025 programme. (Spacecraft: ESA/ATG medialab; Jupiter: NASA/ESA/J. Nichols (University of Leicester); Ganymede: NASA/JPL; Io: NASA/JPL/University of Arizona; Callisto and Europa: NASA/JPL/DLR)

flybys using a process known as gravity assists. After being placed in a heliocentric (Sun centred) orbit, JUICE will return to Earth in May of 2023 and fly onto Venus in October of the same year, picking up more energy in the process. JUICE will return to Earth a second time in September 2024, then flyby Mars in February of the following year. The final gravity assist will be a flyby of Earth in November 2026, followed by a long three year traverse to Jupiter. It is expected that JUICE will arrive in orbit around Jupiter in October 2029. Whilst observations of Jupiter will occur, the large Galilean satellites Europa, Callisto and Ganymede are the main focus of the mission. Several flybys of these icy worlds will set up the orbit of JUICE to approach and finally to go into orbit around Ganymede in 2032 for a mission expected to last around two Earth years. There will be more of JUICE in future editions of the *Yearbook of Astronomy*. In the meantime additional information can be obtained from the mission website at **sci.esa.int/web/juice**

And Briefly…

Commercial Lunar Payload Services
The NASA instigated Commercial Lunar Payload Services programme is calling on private industry to design, develop and launch low cost missions to the surface of the Moon in support of the upcoming Artemis program to return humans to the

Moon. As this is an active and rapidly changing programme, interested readers are encouraged to visit the mission website at **nasa.gov/content/commercial-lunar-payload-services**

Parker Solar Probe: Getting Closer, Getting Hotter

The American Parker Solar Probe – launched on 12 August 2018 – continues its long journey around the Sun for a series of very close passes. The mission is using flybys or gravity assists of Venus to rob the spacecraft of heliocentric velocity so that the next perihelion or closest point to the Sun gets ever closer. During 2022 it will reach perihelion four times. Over the next three years the gravity assists will push the perihelion distance down to within 6.9 million kilometres from the Sun's surface, testing its special heat shield to the limit. Interested readers can follow the journey of the Parker Solar Probe using the mission website **parkersolarprobe.jhuapl.edu**

BepiColombo: Closing in on Mercury

The European BepiColombo mission to Mercury is also using gravity assists to gradually circularise its orbit close to that of Mercury. The spacecraft has already made one flyby of Earth and two of Venus. In 2021 the first of many flybys of Mercury occurred, the second taking place in 2022. By early 2025 the sixth Mercury flyby will almost circularise the orbit of BepiColombo leading to orbit insertion in December of that year. The BepiColombo mission website can be found at **sci.esa.int/bepicolombo**

Solar Orbiter: ESA Explores the Sun

The European Space Agency's Solar Orbiter was launched on 10 February 2020. The mission will provide close-up, high-latitude observations of the Sun. Solar Orbiter will have a highly elliptic orbit between 1.2 au at aphelion and 0.28 au at perihelion. It will reach its operational orbit just under two years after launch by using gravity assists at Earth and Venus. Further gravity assists at Venus will increase its orbital inclination, reaching up to 24° at the end of the nominal mission (approximately seven years after launch) and up to 33° in any extended mission. For further information you can check out the mission website **sci.esa.int/web/solar-orbiter**

OSIRIS-REx: Bringing Home the Bacon

Launched on 8 September 2016, NASA's sample return mission to asteroid 101955 Bennu arrived at the asteroid at the end of December 2018. A complete survey of this 500 metre diameter carbonaceous asteroid was completed over the next

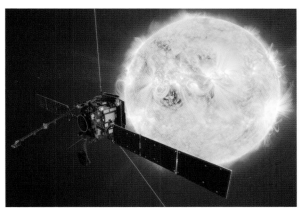

Artist's impression of ESA's Solar Orbiter spacecraft, which will observe our Sun from within the orbit of Mercury at its closest approach. (ESA/ATG medialab)

18 months. The primary aim of the mission was to collect a small sample of this pristine material. This was successfully completed on 20 October 2020 and stowed in the sample return capsule. Departure from Bennu occurred successfully on 10 May 2021. The capsule is due to land at the Utah Test and Training Range in the USA on 24 September 2023. Progress of this sample return mission can be followed at the mission website **asteroidmission.org**

Juno Mission Extended

The Juno mission to Jupiter arrived at the gas giant on 4 July 2016 and was placed into a polar orbit that brings it back to perijove – or closest point to Jupiter – every 53 days. Originally intended to end in 2021, the mission has been granted an extension through to 2025 to allow for an additional 42 orbits, which will provide further observations of Jupiter's north polar regions; flybys of Europa, Ganymede, and Io; and the first extensive exploration of the faint rings encircling the planet, first seen by the Voyager 1 spacecraft in 1979. The Juno mission website can be found at **missionjuno.swri.edu**

As always, Solar System exploration continues to excite and inspire, and next year promises to be no different.

Anniversaries in 2023

Neil Haggath

Nicolaus Copernicus (1473–1543)

One of the most important steps in the history of science – that of realizing that the Earth is not the centre of the Universe – was taken by a man born 550 years ago. (The Greek astronomer Aristarchus proposed the same as early as 280 BC, but was not taken seriously and his work was lost for many centuries.)

Mikołaj Kopernik – always known by his Latinised name of Nicolaus Copernicus – was born on 19 February 1473 in Toruń, in what is now Poland. He studied mathematics and astronomy at the University of Krakow, but later went to Bologna in Italy, where he gained a doctorate in canon law. After

Nicolaus Copernicus: anonymous portrait c.1580, kept in Toruń Town Hall, Poland. (Wikimedia Commons)

further studies in Italy, he returned to Poland at the age of 30 and took up a career in the Catholic Church – though it is unclear whether he was ever ordained as a priest. Around 1512, he settled in the town of Frombork, where he spent the rest of his life.

Astronomy, however, remained his passion. He came to have grave doubts about the Ptolemaic model of the Earth-centred Universe, which for centuries had been incorporated into religious dogma, and proposed that the observed motions of the planets could be better explained by assuming that they, including the Earth, orbit the Sun. His great book, *De Revolutionibus Orbium Coelestium* (On the Revolution of the Celestial Spheres), was completed around 1533, but was not published until a decade later, shortly before his death. Copernicus died on 24 May 1543, aged 70, and is buried in Frombork cathedral.

One of the popular myths of the history of astronomy is that Copernicus dared not publish his book until he was on his deathbed, for fear of reprisals by the Church. This is in fact nonsense. The book did appear in print as he lay dying, but only because he suffered a stroke while it was at the printer's. The truth is that he delayed publication because he was not entirely satisfied with his own work; he knew that there were discrepancies which his model could not explain – they were resolved decades later by Kepler, who realized that the planets' orbits are elliptical, rather than circular – and circulated the manuscript among his friends for their ideas. His work was known for several years, among both astronomers and clergy, before it went into print.

After his death, however, some senior clergy did denounce heliocentrism on theological grounds. In 1616, the Church insisted that a number of "corrections" be made to *De Revolutionibus*; the uncensored version was placed on the Index of prohibited books, where it remained until 1835.

Ejnar Hertzsprung (1873–1967)

150 years ago, two very prominent astronomers were born one day apart. The first was Ejnar Hertzsprung, born in Copenhagen, Denmark on 8 October 1873.

After working at Göttingen and Potsdam Observatories in Germany, he moved to Leiden Observatory in the Netherlands in 1909, where he would serve for the rest of his career, becoming its Director in 1937.

In 1913, Hertzsprung measured the distances of several Cepheid variable stars by parallax, and thus calibrated the Period-Luminosity Relationship discovered by Henrietta Leavitt (1868–1921) the previous year. He subsequently determined the distance of the Small Magellanic Cloud.

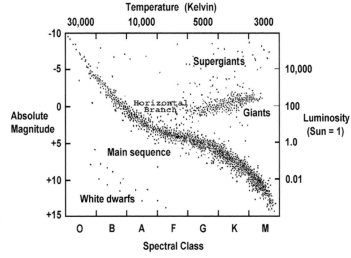

The Hertzsprung-Russell Diagram. (NASA/CXC/SAO) or (NASA/Chandra X-ray Center/Smithsonian Astrophysical Observatory)

He is best known, however, for discovering a relationship between the luminosities of stars and their colours, which indicate their temperatures. He published this work in 1906, in an obscure German journal. Seven years later, Henry Norris Russell (1877–1957), who was unaware of Hertzsprung's work, independently discovered the same relationship. They are given the joint credit for what is now called the Hertzsprung-Russell Diagram, a vitally important tool in astrophysics.

After retiring in 1946, Hertzsprung returned to Denmark. He died on 21 October 1967, aged 94.

Karl Schwarzschild (1873–1916)

We think of black holes as a pretty modern concept – they were not confirmed to exist until the 1970s – but their existence was first postulated over a century ago, by the second of our 1873 duo. Karl Schwarzschild was born on 9 October 1873 in Frankfurt, Germany. Unlike that of his contemporary Hertzsprung – with whom he would work at Göttingen – his life was tragically cut short.

Schwarzschild was a child prodigy; he published two theoretical papers on celestial mechanics by the age of 16! At only 27, he became a Professor at the University of Göttingen; in 1909, he became Director of Potsdam Observatory, then the most prestigious astronomical post in Germany.

Karl Schwarzschild. (Wikimedia Commons)

When the First World War began, he volunteered for military service, and became an army officer; he was posted to the Eastern Front, where he calculated trajectories for long-range artillery shells.

While serving in the army, he worked on General Relativity, and produced the first exact solutions of Einstein's Field Equations; this impressed Einstein himself, who had only found approximations. As part of this work, he realized that a hypothetical body of large mass and extremely small size would have an escape velocity greater that of light – what we now call a black hole. He postulated that dead stars might collapse under their own gravity to such a state; if such an object collapsed to a radius of $R_S = 2GM/c^2$ (where M is its mass, G the Universal Constant of Gravitation and c the velocity of light in vacuum), its escape velocity would equal that of light. This is now called the Schwarzschild Radius, and marks the event horizon of a black hole, from inside which nothing can escape.

Like so many soldiers of that war, Schwarzschild was killed not by the fighting, but by disease; he contracted a fatal skin disease, and died on 11 May 1916, aged only 42.

Edward Emerson Barnard (1857–1923)

One of the greatest visual observers who ever lived died a century ago. Edward Emerson Barnard was born in Nashville, Tennessee on 16 December 1857. He was raised in poverty, his father having died before he was born; he had barely any formal education, and had to go to work, as a photographer's assistant, at the age of nine!

He developed a passion for astronomy, and bought a 5-inch refractor at age 18. Five years later, he discovered his first comet. When he married in 1881, money was still desperately short, but astronomy came to his rescue. A wealthy businessman announced a prize of $200 for discoveries of comets. Barnard claimed five such

Edward Emerson Barnard.
(Wikimedia Commons/BILD-PD)

prizes, and used the money to build a house; he always said that his house was "built on comets". He discovered 15 comets in all, and became renowned for his exceptional eyesight.

He became well-known in Nashville, and a group of local amateur astronomers funded a scholarship for him at Vanderbilt University. In 1887, Barnard joined Lick Observatory in California. In 1892, using the 36-inch refractor, he discovered Amalthea, the first satellite of Jupiter to be discovered since the four Galileans, and the last planetary satellite to be discovered visually.

In 1895, he joined the new Yerkes Observatory in Wisconsin, where two years later, the world's biggest refractor was completed. He worked there for the remainder of his life. During extensive studies of Mars, he was unable to see Percival Lowell's supposed "canals", and correctly deduced that they do not exist. He did, however, claim to have observed craters on the planet – but did not publish his observations for fear of ridicule. John Mellish also claimed to see Martian craters in 1917, but they were not confirmed to exist until 1965, when the Mariner 4 probe flew by the planet.

In 1916, Barnard found that a faint red star in Ophiuchus has the greatest proper motion of any star. This star, the closest to the Sun after the Alpha Centauri system, is now named Barnard's Star. He is also known for his catalogue of 182 dark nebulae, first published in a paper in the *Astrophysical Journal* in 1919.

Barnard died at Williams Bay, Wisconsin on 6 February 1923. He is buried in his home town of Nashville.

Gerard Kuiper (1905–73)

One of the pioneers of modern planetary science died 50 years ago. Gerrit Pieter Kuiper was born in Holland on 7 December 1905. He graduated from Leiden University, where he was a student of Ejnar Hertzsprung (see above).

In 1935, he went to Yerkes Observatory. After marrying an American woman, he became an American citizen in 1937, and anglicized his given names to Gerard Peter. From 1947 to 1949, he was Director of McDonald Observatory, Texas. He specialized in studies of the Moon and planets; he discovered Uranus' tiny satellite Miranda and Neptune's Nereid.

Gerard Kuiper. (Wikimedia Commons/NASA)

In 1960, Kuiper founded the Lunar and Planetary Laboratory in Tucson, Arizona; he served as its Director for the rest of his life, and helped to identify suitable landing sites for the Apollo programme.

Today, Kuiper's name is most commonly associated with something which he did not quite predict! Kenneth Edgeworth (1880-1972) proposed in 1938 that Pluto might be only one of a large population of small bodies beyond the orbit of Neptune. Kuiper later postulated that such a population had existed early in the Solar System's history, but would have been gravitationally dispersed, and would no longer exist. In the latter part, he was proved wrong in 1992, with the discovery of the first trans-Neptunian object other than Pluto and Charon. Today, thousands are known, in what is now called the Edgeworth-Kuiper Belt.

Kuiper died on 23 December 1973, while on holiday in Mexico. He is one of only three astronomers – the others being Giovanni Schiaparelli and Eugenios Antoniadi – to be honoured with named features on the Moon, Mars and Mercury.

Skylab

50 years ago, NASA launched its first space station – the only one to be operated exclusively by the United States. It was occupied by three successive three-man crews between May 1973 and February 1974, for durations of 28, 56 and 84 days.

Skylab was built from leftover hardware from the Apollo programme. The hull of its main Orbital Workshop component was that of an S-IVB rocket stage; it was launched by the first two stages of a Saturn V rocket, with the station in place of the third stage. The crews were ferried to and from the station by Apollo spacecraft, launched by smaller Saturn IB rockets.

Skylab, photographed by its departing final crew. (Wikimedia Commons / NASA)

The first two crews were commanded by Charles "Pete" Conrad (1930–99) and Alan Bean (1932–2018), who had both landed on the Moon on Apollo 12; the remaining crew members were all first-timers. Each crew also included one scientist-astronaut.

Skylab was launched on 14 May 1973, by the last Saturn V to fly. At 82 feet long and 76 tons, it remains the largest single payload ever launched into orbit. But during the launch, disaster struck; the station's micrometeorite and thermal shield was torn off, and with it one of the two main solar panel "wings". After it reached orbit, the second solar panel jammed and failed to deploy. Skylab was left without power or protection against overheating.

The first crew launch, scheduled the following day, was delayed for ten days, while engineers worked out how the station could be repaired. A "sunshade" was improvised from mylar sheeting and fishing rods, and was loaded aboard the ferry vehicle only a few hours before launch. On reaching the station, Conrad's crew had to perform two dangerous spacewalks, to install the makeshift thermal shield and free the jammed solar panel. The repairs were successful.

The other two missions were hugely successful and productive, with the crews carrying out 270 experiments in life sciences, solar physics, astrophysics, Earth observation and materials processing.

Skylab was de-orbited and re-entered the atmosphere on 11 July 1979. This was intended to happen over the Indian Ocean, but several large pieces fell onto the Outback of Western Australia.

Another "first" was that the missions had a rescue capability. Skylab had two docking ports; a modified Apollo spacecraft was built, which could carry five men, and a two-man crew trained to fly it. Had any of the crews had a problem with their own Apollo, the rescue vehicle would have been used to bring them back.

Betelgeuse

Tracie Heywood

Although Betelgeuse may be one of the most famous stars in the night sky, it is not the brightest star in Orion. Magnitude 0.1 Rigel is brighter. For some reason, the German astronomer Johann Bayer allocated the designations Alpha (α) Orionis to Betelgeuse and Beta (β) Orionis to Rigel when assigning Greek letters to the stars of Orion in 1603. Given that Betelgeuse does vary in brightness, some people have suggested that at the time Betelgeuse may have been brighter than Rigel, but there is no evidence to support this.

The magnitude range of Betelgeuse is usually listed as being +0.4 to +1.3 although, as has been seen in recent years, Betelgeuse can drop below the fainter

Image of Orion captured in the early hours of 8 Sep 2021. Note how the relative brightness of the stars at lower altitudes is impacted by haze and the glare from street lighting. (Paul Sutherland/Society for Popular Astronomy)

limit. Monitoring the brightness changes of this star can be challenging. There are not many other stars as bright as Betelgeuse and all of them are far enough away on the sky for their apparent brightness to be affected differently by haze and by the glare from street lighting. My personal experience has shown that Betelgeuse spends most of its time varying between magnitudes +0.5 and +0.9. This brightness

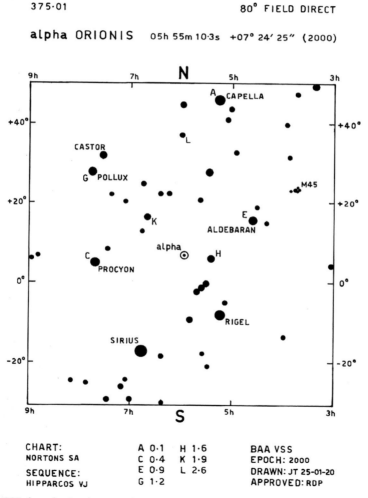

The BAA VSS chart for Betelgeuse (Alpha Orionis), showing comparison stars (labelled A to L) with which its brightness can be compared. (BAA Variable Star Section)

Light Curve for ALPHA ORI

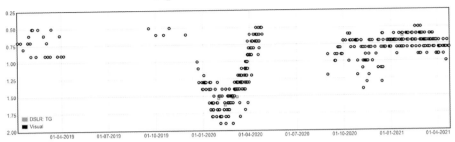

Betelgeuse Light Curve (2019–2021)
The BAA VSS light curve for Betelgeuse from January 2019 to April 2021. (BAA Variable Star Section)

range is quite nicely bracketed by the brighter stars Procyon (magnitude +0.5) and the fainter Aldebaran (magnitude +0.9).

During the first couple of hours after Orion rises Aldebaran is somewhat higher in the sky than Betelgeuse while Procyon is lower down. Consequently, any haze that may be present may cause Betelgeuse to appear brighter than Procyon and fainter than Aldebaran! For this reason, I was initially quite sceptical when I heard reports during the autumn of 2019 that Betelgeuse had started to become unusually faint. As soon as clear skies allowed me to observe Betelgeuse, however, I could clearly see that a deep fade had indeed occurred.

Some people speculated/hoped that this unusual fading might be a pre-cursor to Betelgeuse going supernova. It is known that Betelgeuse is a massive star that will almost certainly end its life as a type II (core collapse) supernova. There is a problem, however, in that astronomers do not know the exact sequence of visible events that occur prior to a massive star exploding as a supernova. The most recent supernova observed in our own galaxy was Kepler's Supernova of 1604, just prior to the invention of the telescope. For many decades, it was assumed that type II supernovae occurred in red supergiant stars (like Betelgeuse), but when astronomers searched old images for the precursor of the 1987 type II supernova in the Large Magellanic Cloud, they found that it had been a blue supergiant star. Historical images had also shown no indication of variability prior to the supernova.

As we now know, Betelgeuse reached a minimum at about magnitude +1.6 in February 2020 and then started to brighten again. By the time that it was lost in the evening twilight in May 2020, it was more or less back to normal. Another less shallow minimum – down to magnitude +1.0 – occurred in August 2020, the 2020-2021 apparition closing with a shallower minimum in April 2021.

Why had the early 2020 fade been so deep? Historical observations did record spells with deep minima during 1945-48, 1952-53, 1978-80 and 1984-1986, so it was clear that the recent deep fade, though dramatic and better observed, was not wholly unprecedented. Much about Betelgeuse – including its precise mass, size and distance – was only vaguely known and, although books often stated a 2,070 day period for its variations, many alternative periods of variation have been suggested.

For the reasons outlined earlier, the scatter in visual observations can be large. In addition, for many decades the photometers typically used by professional astronomers would become saturated when attempting to measure the brightness of Betelgeuse. Betelgeuse is also too distant for accurate trigonometric parallax measurements to be made from the Earth's surface. Any uncertainty regarding a star's brightness and distance result in estimates of its mass, size and 'life expectancy' being less reliable. Consequently, there was uncertainty as to how bright Betelgeuse would shine as a supernova. Would its peak brightness be comparable with that of a Full Moon (magnitude −12.5) or 'only' as bright as bright as the First Quarter Moon (magnitude −9)?

More recently, Hipparcos and other space missions have provided more certainty over its distance (although not all agree). The consensus, however, seems to be that Betelgeuse lies at a distance of about 550 light years and has a diameter around 750 times that of our Sun (making it larger than the size of Jupiter's orbit). In addition, satellites such as Coriolis – together with improved ground-based photometry – have tied down its brightness variations better. There appear to be periods of approximately 185 days, 420 days and 6.5 years related to pulsations in the outer layers of Betelgeuse. Additional variations may be due to the rotation of Betelgeuse (believed to take approximately 35 years); giant starspots (related to huge convection cells); and dimming by clouds of 'dust' that can be ejected into the surrounding space from the tenuous outer layers of Betelgeuse. It is this latter cause – in conjunction with one of the pulsation minima – that is believed to have been responsible for the deep minimum of early-2020.

The pulsation periods are indicative of Betelgeuse having a mass of around 18 times that of our Sun and early in the phase of its evolution during which deep down inside it is 'burning' helium into carbon. This would suggest that Betelgeuse may be as much as half a million years away from going supernova, and that we are not about to 'lose' this famous star!

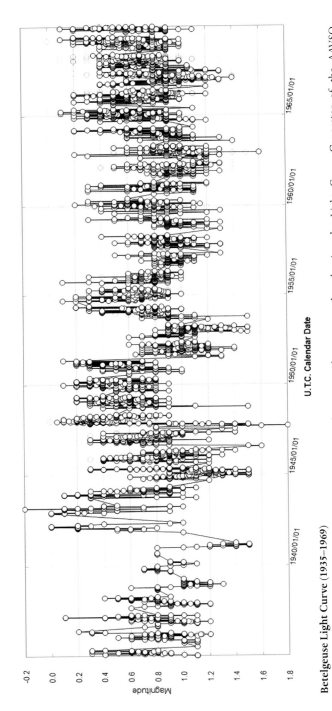

Betelgeuse Light Curve (1935–1969)
Visual light curve for Betelgeuse from 1935 to 1969, showing 20-day means, plotted using the Light Curve Generator of the AAVSO. (AAVSO International Database **www.aavso.org**)

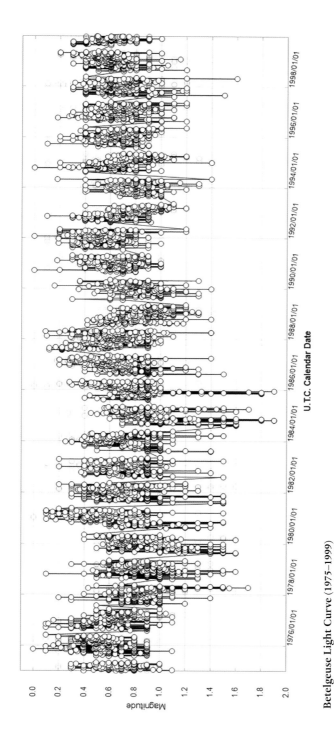

Betelgeuse Light Curve (1975–1999)

Visual light curve for Betelgeuse from 1975 to 1999, showing 20-day means, plotted using the Light Curve Generator of the AAVSO. (AAVSO International Database **www.aavso.org**)

Optical SETI at Harvard

David M. Harland

When the search for extraterrestrial intelligence (SETI) kicked off in 1960 it sought radio signals because radio was the only technology available to us at that time which was capable (albeit only just) of spanning interstellar distances. However, in that year the first laser (Light Amplification by Stimulated Emission of Radiation) was created by Theodore H. Maiman at Hughes Research Laboratories,[1] exploiting the concepts of Charles H. Townes at Columbia University and Arthur L. Schawlow at Bell Laboratories.[2] In 1961 Townes and Robert N. Schwartz, also at Columbia, suggested that aliens might communicate across interstellar distances using lasers.[3]

In 1971 NASA funded the Project Cyclops summer study at Stanford University to review the status of radio searches and define future ambitions.[4] It discussed the possibility of searching for lasers (naming it optical SETI) but ruled it out as not yet feasible.

An optical SETI project faced an early problem of perception, in that Schwartz and Townes had presumed that aliens would use continuous-wave lasers, reasoning that unless the laser drew an inordinate power it would be lost in the glare of the star of the alien planet.

In fact, Monte Ross, who led the development of the first space laser communications system by McDonnell Douglas in St. Louis, Missouri, had pointed out in 1965 that aliens were more likely to use short-pulse lasers.[5] Indeed, nanosecond-scale pulses would not only provide a high peak power at the source without imposing an excessive average power rating on the transmitter, the number of photons that are received from the laser could far outshine the parent star during the time of the pulse.

1. 'Stimulated Optical Radiation in Ruby', T. H., Maiman, *Nature*, **187**, 493–494, 1960.
2. 'Infrared and Optical Maser', A. L., Schawlow and C. H. Townes, *Physical Review*, **112**, 1940, 1958.
3. 'Interstellar and Interplanetary Communication by Optical Masers', R. N. Schwartz and C. H. Townes, *Nature*, **190**, 205–208, 1961.
4. 'Project Cyclops: A Design Study of a System for Detecting Extraterrestrial Intelligent Life', Stanford University with NASA-Ames, NASA-CR-114445, 1971.
5. 'Search Laser Receivers for Interstellar Communications', Monte Ross, Proceedings IEEE, **53**, 1780, 1965.

Furthermore, whereas receivers for radio SETI were fine-tuned to the frequency of the transmission, for optical SETI we would not require to know the frequency of a pulsing laser. And the fact that a laser shines at a single frequency while a star has a broad spectrum meant an optical receiver would be able to reject much of the spectrum and still have a much wider spectral acceptance than a radio receiver. All that would be required would be to detect the fluctuating intensity.

Optical SETI does not employ the same method of detection as for radio SETI, namely 'coherent' detection (also known as 'heterodyne'); it is instead possible to employ 'direct' detection, where the phase of the carrier frequency of the laser is discarded and only the amplitude of the pulses is necessary. It is therefore possible to construct large collectors more cheaply than similarly sized ones intended for imaging purposes whose 'figure' must preserve the phase of the light across the width of the collector.

Another benefit of optical SETI was that 'noise' would be simpler to overcome. Radio SETI has to contend with interference from terrestrial sources. And, of course, there is the noise which is intrinsic to the receiver itself. Although the most sensitive radio detectors were chilled down to almost absolute zero to minimise internal noise, it could not be eliminated entirely. Due to the differences in noise at radio and optical frequencies, direct detection for optical SETI would be able to achieve a sensitivity that approached coherent detection without the complexities and limitations of the latter.

Hence, Ross argued, a pulsed laser would be a more sensible means of undertaking interstellar communication than either radio frequencies or continuous-wave lasers.

And from the point of view of mounting a search, we could be fairly sure that optical pulses on the nanosecond scale were of artificial origin.

In communicating across interstellar distances it is beneficial to use a narrow beam to direct the maximum energy toward the target. This is more readily achieved using a laser than a radio transmission.

For a target star of a given spectral type, it would be ideal to scale the beam so that it spans the diameter of the 'habitable zone' of that star on arrival. But the farther the star and the narrower the beam, the greater is the difficulty in confining the beam to that zone. In fact, the attainable beam width measured in radians is approximately the wavelength of the transmission divided by the radiating antenna's diameter. As the wavelength of visible light is six orders of magnitude shorter than microwaves, it is evident that wavelengths shorter than, at, or near the visible-light range are required in order to make a beam that spans the habitable zone of a remote star.

Seeking stars hosting life capable of benefiting from a transmission, an alien civilisation might work through a list of candidates, providing a pulsed signal as a 'beacon' to attract attention, and follow this up with a modulated laser that supplies the bulk of the information on offer.

The desire to maximise the energy delivered by a laser at interstellar distances by narrowing its beam to match the diameter of the habitable zone of a star could easily be undermined by an inability to accurately *aim* the beam. Even employing excellent hardware which can limit the raw misalignment and the dynamic jitter to an overall pointing error no greater than 10% of the beam width, there remains a potentially greater uncertainty in pointing arising from uncertainty in the target's precise location on the sky.

Stars pursue independent motions in space, with some travelling in more or less circular orbits in the disk of the galaxy that we know as the Milky Way while others move in elliptical orbits set at all angles to that plane. Even at the speed of light, a signal might take centuries to reach its target. If the laser were to be aimed precisely at the star in the sky, when the beam attains that distance the star will have moved. Both the speed and direction of a star must be known very accurately in order to calculate how much to offset the laser from the current location of the star on the sky. This 'point-ahead' calculation will be different for each target. Although achieving this may be difficult, any civilization that is using lasers for interstellar communication must have figured out how to do it. As, indeed, will we, if ever we receive a 'cold call' from an alien civilisation and decide to send a reply in the same manner.

In its simplest form, direct detection for optical SETI uses a telescope, usually referred to as a 'light bucket', equipped with a suitable detector such as a photomultiplier.

A photomultiplier is an electronic apparatus which converts incident photons into electrons, using successive 'cascade stages' to amplify the number of electrons until the signal is above the electrical noise. A later technology exploits silicon or gallium arsenide to achieve the same goal of high-speed detection of individual photons. But even using excellent hardware there is a fundamental uncertainty inherent in the detection of a signal, known as 'quantum noise' or 'photon noise'.

For radio SETI, the finely-tuned detector has to lock on to the signal and accommodate shifts in frequency caused by the motion of an alien planet around its host star, or of the star relative to Earth. However, the spectral response of the light-sensitive portion of a photomultiplier is much broader than any likely Doppler shift, so this complication does not arise.

Several experiments in the early 1990s inspected a selection of stars for laser pulses, most notably by Stuart A. Kingsley of Columbus, Ohio, who did so as a private venture using a telescope with a diameter of 10 inches, and by Guillermo A. Lemarchand using a 2.15-metre telescope at the El Leoncito Astronomical Complex in the San Juan Province of Argentina, with neither detecting any signals.[6, 7]

In 1998 Paul Horowitz at Harvard University led a team which built a custom-designed nanosecond detection system. This was installed on the 1.55-metre telescope at the nearby Oak Ridge Observatory, operated by the Harvard-Smithsonian Center for Astrophysics. After a successful trial, a similar instrument was installed on the 0.9-metre telescope at the Fitz-Randolph Observatory in New Jersey by David T. Wilkinson of Princeton University. The two telescopes then undertook simultaneous observations of the same target stars. Operating a pair of telescopes far apart eliminated 'false positives' arising from local phenomena such as atmospheric Čerenkov flashes, cosmic ray muons, or electron showers.

The Targeted Optical SETI Instrument on the day of its "first light" with the 1.55-metre telescope at the Oak Ridge Observatory, operated by the Harvard-Smithsonian Center for Astrophysics. L-R: Robert Stefanik, Joe Zajac, Costas Papaliolios, Chip Coldwell, and Jonathan Wolff. (Paul Horowitz)

In 2,378 hours of operations between October 1998 and November 2003 the teams accumulated a total of 15,987 observations of 6,176 stars, typically between 2 and 40 minutes per session. The targets included all of the main sequence dwarf stars between spectral types A and early-M within 100 parsecs (326 light years) of

---

6. 'The Search for Extraterrestrial Intelligence (SETI) in the Optical Spectrum: A Review', Stuart A. Kingsley, SPIE conference 'The Search for Extraterrestrial Intelligence (SETI) in the Optical Spectrum', **1867**, 75–113, S. A. Kingsley (Ed.), 1993.

7. 'Interplanetary and Interstellar Optical Communications Between Intelligent Beings: A Historical Approach', Guillermo A. Lemarchand, SPIE conference 'The Search for Extraterrestrial Intelligence (SETI) in the Optical Spectrum II', **2704**, S. A. Kingsley and G. A. Lemarchand (Eds), 1996.

The roll-off-roof observatory for the 1.83-metre Optical SETI telescope which operates on a meridian mount. L-R: Jason Gallicchio and Curtis Mead. (Paul Horowitz)

us and between Declinations −20° and +60°, the part of the celestial sphere visible from those two sites. No alien signals were detected.[8]

The Harvard-Princeton experiment showed it was essential that the system be robust against many possible failure modes, and that it have reliable software and good telemetry for virtually automatic operation with low maintenance. One important lesson was learned in June 2001, when the wire cables between the control room and the telescope of the Harvard system acted as antennas in what the team called "the mother of all thunderstorms" and destroyed the electronics connected at both ends. All the wire cabling was promptly replaced with optical fibre as protection against future lightning storms.

These insights were applied by Horowitz's team in their follow-on project at the Oak Ridge Observatory. This time, rather than examining specific stars, the plan was to survey the sky. For this, they developed a new telescope and a new instrument.[9, 10, 11] Since the phase information was not required in using direct

---

8. 'Search for nanosecond optical pulses from nearby solar-type stars', A. W. Howard, P. Horowitz, D. T. Wilkinson, C. M. Coldwell, E. J. Groth, N. Jarosik, D. W. Latham, R. P. Stefanik, A. J. Willman, J. Wolff J and J. M. Zajac, *Astrophysical Journal*, **613**, 1270–1284, 2004.

9. 'An All-Sky Optical SETI Survey', Andrew Howard, Paul Horowitz and Charles Coldwell, American Institute of Aeronautics and Astronautics, Inc., 2000.

10. 'Targeted and All-Sky search for Nanosecond Optical Pulses at Harvard-Smithsonian', P. Horowitz, C. Coldwell, A. Howard, D. Latham, R. Stefanik, J. Wolff and J. Zajac, *Proceedings SPIE*, **4273**, 119–127, 2001.

11. 'Astronomical Searches for Nanosecond Optical Pulses', Andrew Howard (thesis), 2006, **seti.harvard.edu/grad/apdf/howard_thesis_compact.pdf** (8 megabytes).

The 70-kg All-Sky Camera (with its covers off) at the time of its completion, very late one night. L-R: Paul Horowitz, Andrew Howard, Jason Gallicchio, Steve Howard, and Curtis Mead. (Paul Horowitz)

The f/2.5 Optical SETI telescope. The 0.9-metre secondary is angled in order to direct the light into the All-Sky Camera mounted on the side of the telescope (covers were installed in operation). Present are Andrew Howard and (foreground) Curtis Mead. (Paul Horowitz)

At the computers for "first light" of the All-Sky Camera. L-R: Steve Howard, Curtis Mead, and Jason Gallicchio. (Paul Horowitz)

detection to look for nanosecond-scale laser pulses, it was possible to use less accurately figured optics. This meant that their custom-designed light bucket, which had a resolution of only 2 arc minutes, no sharper than the human eye, was considerably cheaper than a telescope of comparable size intended for imaging.

The primary mirror of the f/2.5 telescope was 1.83 metres in diameter and, unusually, the secondary was 0.9 metres in diameter and angled to direct the light into an instrument on the frame alongside the telescope, rather than behind it.

Whereas the earlier system had used a photon detector with a single aperture that received light directly on the optical axis, this time they used an array of 16 'pixelated' photomultipliers, together comprising a grid of 1,024 apertures, to simultaneously inspect a rectangular patch of sky.

As a further cost-saving measure, the telescope was mounted in the transit style, meaning that it maintained the same azimuth and moved only in altitude. The field of view was 0.2° in Right Ascension and 1.6° in Declination. Once set at a certain altitude, the Earth's rotation would cause the field of view to scan in Right Ascension. In effect, one night's observations would fill in a slice of the sky. On successive nights, the altitude would be adjusted to obtain the next slice, progressively going north or south. It would take a year to fill in the entire sky visible from that site. The fact that the telescope was fixed in azimuth meant it

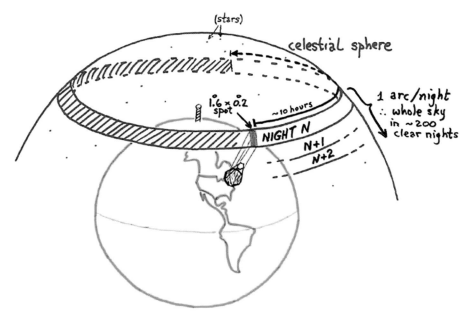

Being a meridian telescope, the all-sky survey was assembled in strips by selecting a succession of Declination angles and allowing the rotation of the Earth to scan in Right Ascension. (Paul Horowitz)

could be housed in a wooden shed with a roll-off roof. The building, telescope and instruments were all designed to be commanded from afar.

With 1,024 channels simultaneously providing data on a nanosecond scale, the signal processing system developed by Andrew Howard, one of Horowitz's graduate students, had to handle a vast amount of data. Its 32 custom-designed chips could run at a rate of 3.5 trillion bits per second, seeking a large spike in the photon count that might be a laser pulse. The detectors were split into two arrays, so that if one array were to detect an interesting signal, this could be checked against the other array. A pulse was accepted only if it was detected by both channels in the same nanosecond.

One key difference between targeted and all-sky data related to the definition of an 'observation'. In the targeted search, observations were of given stars for fixed durations, but all-sky observations were to be of the sky. In fact, as the sky drifted across the pixelated photomultiplier, each point source was examined for about 1 minute. The all-sky search therefore would not record the names of the stars it observed; it sensed the whole northern sky. The source of a positive detection

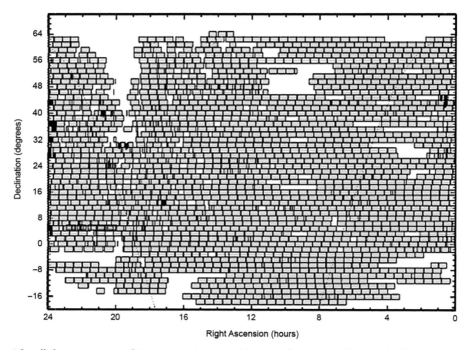

The all-sky survey as work-in-progress in 2007 with 2,217 observations for a total of 992 hours. (Curtis Mead)

would have to be calculated from the pointing of the telescope and the timing of the signal on the grid of pixels.

The Harvard All-Sky Optical SETI Survey started in 2006, with most of the up-front funding donated by The Planetary Society. Over several years, the sky in the designated Declination range was scanned several times. After false positives were eliminated there was no evidence of an alien laser.[12, 13]

As a follow-on project, the Advanced All-sky Camera was developed by Curtis Mead, another of Horowitz's students, to overcome some of the technical

12. 'Challenges in the first all-sky optical SETI', C. Mead, P. Horowitz, A. W. Howard, P. Sreetharan, J. Gallicchio, S. Howard, C. Coldwell, J. Zajac and A. Sliski, *AIAA 57th International Astronautical Congress, IAC 2006*, **3**, 1635–1642, 2006.

13. 'Initial results from Harvard all-sky optical SETI', A. Howard, P. Horowitz, C. Mead, P. Sreetharan, J. Gallicchio, S. Howard, C. Coldwell, J. Zajac and A. Sliski, *Acta Astronautica*, **61**, 78–87, 2007.

limitations of its predecessor. Observations using the same telescope and much of the existing software began in September 2012.[14]

As always, the lessons learned from constructing and operating one system suggest improvements for future instruments. In particular, the progression of all-sky surveys should tend towards larger light buckets and more photomultiplier pixels. A larger collector increases sensitivity by lowering the number of photons per square metre needed to detect a distant laser pulse. This permits detection of smaller pulse energies and more distant transmitters. Additional pixels increase the amount of sky that is observable, reducing the time for a full survey and increasing the rate at which the process can be repeated.

Harvard is also involved in a proposal to create a considerably more sophisticated Pulsed All-sky Near-infrared Optical SETI (PANOSETI) system. This dedicated SETI facility will seek to increase the area of sky searched, the range of wavelengths covered, the number of star systems observed, and the duration during which they are monitored. This will be enabled by the rapid technological advance of fast-response optical and near-infrared detector arrays. Furthermore, the instrument will use a pair of innovative domes, each equipped with 45 Fresnel lenses to search simultaneously for transient signals over 4,500 square degrees of sky (duplicated). A 1 kilometre separation between the domes will facilitate rejection of local transients and discrimination of Čerenkov atmospheric events through parallax. If there is an alien civilisation flashing a laser at us, we will stand a good chance of finding it.[15]

For Horowitz the challenge is irresistible because, "While this search is undoubtedly a long shot, the payoff would be profound. It would be without doubt the greatest discovery in human history."

-oOo-

I would like to thank Paul Horowitz of the Optical SETI project at Harvard University for reviewing the manuscript and supplying the illustrations.

14. 'A Configurable Terasample-per-second Imaging System for Optical SETI', Curtis Mead (thesis), 2013, **dash.harvard.edu/handle/1/11158246** (35 megabytes).

15. 'Panoramic Optical and Near-Infrared SETI Instrument: Overall Specifications and Science Program', Shelley A. Wright, Paul Horowitz, Jérôme Maire, Dan Werthimer, Franklin Antonio, Michael Aronson, Sam Chaim-Weismann, Maren Cosens, Frank D. Drake, Andrew W. Howard, Geoffrey W. Marcy, Rick Raddanti, Andrew P. V. Siemion, Remington P. S. Stone, Richard R. Treffers, and Avinash Uttamchandani, Proc. SPIE, **10702**, article. 107025L, 2018.

A Brief History of the End of the Universe

David Harper

Humanity has imagined the end of the universe in many ways: The Apocalypse. Ragnarok. The Last Judgement. Götterdämmerung. Milliways. It has been depicted by artists in literature, paintings and opera. Science can also predict how the universe will end. In this article, I will explore some of the more interesting chapters in the future history of the cosmos as foretold by astronomy and physics.

In the Beginning

To begin at the beginning. About 13.8 billion years ago, the Universe came into existence in an event known as the Big Bang. Theoretical physicists are able to describe the state of the Universe in some detail from a very tiny fraction of a second after the Big Bang. The first elements, hydrogen and helium, formed when

The English artist John Martin vividly imagined the end of the world in the painting "The Great Day of His Wrath" in 1851–3. It is part of his triptych of huge canvasses on the theme of the Last Judgement at the Tate Gallery. (Google Art Project / Tate Britain)

the Universe was only a few minutes old. For the first 400,000 years, the Universe was so hot that it existed as a plasma of photons, electrons and atomic nuclei. As the Universe cooled further, electrons became bound to the atomic nuclei, and space became transparent. By about half a million years, the first stars and galaxies formed. There was light.

The first stars were massive, ultra-luminous and very short-lived, dying in colossal supernova explosions which seeded their galaxies with heavier elements. Later generations of stars further enriched the Universe with the building blocks from which the Earth, and all life on it, would be made: carbon, oxygen, nitrogen, phosphorus and iron.

The Ascent of Humans

Our Sun was born 4.6 billion years ago, and the Earth formed at the same time. There is tentative evidence that life had already become established on Earth nearly 4 billion years ago, almost as soon as the surface had cooled enough for liquid water to exist. For a billion years, life was single-celled. The Earth's primordial atmosphere was a mainly carbon dioxide, ammonia and water vapour. It was devoid of oxygen until about 2.3 billion years ago. More complex multi-cellular forms of life slowly emerged, but it was not until 500 million years ago that a rapid increase in biodiversity known as the Cambrian explosion gave rise to the animal forms that would evolve into the wide variety that we know today. Human beings, our

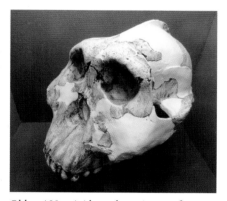

Olduvai Hominid number 5 is one of our oldest-known hominid ancestors. The 1.75 million-year-old fossilized skull was found in 1959 at Olduvai Gorge in Tanzania by the celebrated anthropologist Mary Leakey. (Daderot at Wikimedia Commons / Springfield Science Museum)

own species *Homo sapiens*, emerged only 300,000 years ago. The whole of recorded history goes back no more than 10,000 years. Humans discovered the true size and age of the Universe less than a century ago.

The Big Crunch

One of the most pressing questions in cosmology is whether the Universe will continue to expand forever, or whether the expansion will eventually stop and go into reverse. This possibility was demonstrated by the Russian mathematician

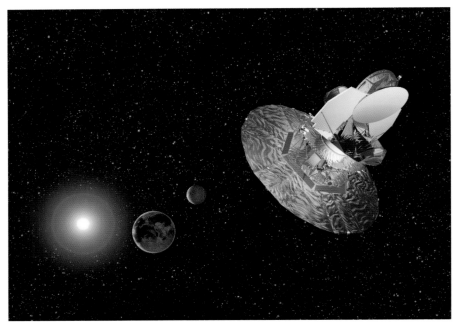

NASA's WMAP spacecraft was launched in 2001 to explore the cosmic microwave background, sometimes called "the afterglow of the Big Bang". This artist's impression shows the spacecraft approaching the Sun-Earth L2 Lagrangian point. (NASA/WMAP Science Team)

Alexander Friedmann in 1922 as a consequence of Einstein's equations of general relativity, provided that the Universe contained enough matter for the force of gravity to overcome the expansion of the Big Bang.

Cosmologists use the symbol Ω to denote the ratio of the overall density of the Universe to the critical density. If Ω is less than 1, then the Universe will continue to expand forever, because there is not enough matter to allow the force of gravity to slow down the expansion. However, if Ω is greater than 1, then gravity will eventually prevail. The Universe will cease to expand at some point in the far distant future, and it will then begin to contract, ending in a cataclysmic event sometimes called the Big Crunch.

Observational evidence from various sources, including the Wilkinson Microwave Anisotropy Probe spacecraft in the early 2000s, yields a value of Ω which is exactly 1, within the margins of error of the data. It appears that our Universe sits on a knife-edge. It will continue to expand forever, but ever more slowly.

The End of Life on Earth

When our Sun began its career as a Main Sequence star 4.6 billion years ago, it was 70% less luminous than it is today. As it consumes its reserves of hydrogen, it will continue to become more luminous by about 10% every billion years. This has serious consequences for life on Earth, because as the Sun's luminosity grows, the Earth receives more heat energy from the Sun. A billion years from now, it seems likely that the surface temperature of the Earth will have risen in response to the Sun's increasing luminosity to the point where liquid water can no longer exist. The Earth's oceans will have evaporated, turning the Earth into a Venus-like world completely covered by clouds. This spells the end of life on our planet, because whilst life can

In 1974, NASA's Mariner 10 visited Venus. Modern image-processing techniques have been applied to the Mariner 10 data to produce this view of the Earth's twin which brings out subtle detail in the upper atmosphere. (NASA/JPL-Caltech)

survive in many extreme environments, from volcanic vents to the interiors of nuclear reactors, biologists agree that the one essential ingredient is liquid water.

As the Sun continues to become brighter and hotter, the water vapour in the Earth's atmosphere will break down into hydrogen (H) and hydroxyl (OH), and the hydrogen atoms will be lost into space. Eventually, the Earth will become a hot, dry, airless world like Mercury.

Don Korycansky and colleagues proposed an intriguing way to avoid this fate in an article published in 2001. They suggested that humanity could divert large asteroids from the Kuiper Belt and swing them past the Earth in a "gravity assist" manoeuvre similar to those that NASA uses to boost the trajectories of spacecraft such as Voyager and New Horizons. In this scenario, each asteroid encounter would increase the Earth's orbital distance from the Sun by a fraction. They calculated that one encounter every 6,000 years should pull the Earth to a higher orbit sufficiently quickly to counteract the increasing luminosity of the Sun. For the moment, however, this idea remains firmly in the realm of science fiction.

The End of the Earth Itself

The Sun is a Main Sequence star. It is burning hydrogen at its core, converting it to helium by nuclear fusion. It has been doing this for 4.6 billion years, and it will continue to do so for another five billion. Eventually, though, the Sun's core will run out of hydrogen. The accumulated helium will collapse under its own gravity, and nuclear fusion will then begin to consume hydrogen in a shell surrounding the helium core. In response, the outer layers of the Sun's atmosphere will expand, and our Sun will become a red giant star. As its diameter increases, it will engulf Mercury and then Venus. Both planets will be vaporised.

There has been much debate among experts on the evolution of the Sun as to whether the Earth will share the same fate. One school of thought says that as the Sun ages, it loses mass via the solar wind, and the orbits of all of the planets will expand in response, allowing Earth to escape engulfment.

Klaus-Peter Schröder and Robert Connon Smith examined the interaction between the Earth and the Sun during its red giant phase in more detail, and reported their findings in a paper in 2008. They agreed that the Earth's orbit will expand as the Sun's mass falls to just 66% of its current value during its red giant phase, but this will not save our planet from a fiery end.

They go on to explain why. As the Sun expands to become a red giant, its rotation will slow down, and eventually stop completely. The Earth will be orbiting just above the surface of the ageing Sun, and it will be raising tides within the upper layers of the Sun's atmosphere, just as the Moon raises tides in the Earth's oceans. This will drain the Earth of its orbital angular momentum, and over less than three million years, the Earth will spiral into the outer atmosphere of the Sun. At this point, gas drag forces will accelerate the process, and our planet will quickly burn up.

The End of the Age of Stars

The Universe is filled with stars. It seems such an obvious fact, at least to astronomers, and yet it was not always so. And in the distant future, it will cease to be so again. We live in a special period, which astronomers have named the Stelliferous Era, or the Age of Stars.

The earliest stars formed just half a million years after the Big Bang. Perhaps counter-intuitively, astronomers call this first generation of stars Population III. They were composed purely of hydrogen and helium, since those were the only elements which existed in the infant Universe. Most of these stars were very massive, extremely luminous, and short-lived, ending in supernova explosions after just a million years and seeding their galaxies with heavier elements such as oxygen, carbon and iron which would be incorporated into later generations of stars.

The Sun is a third-generation star which formed 4.6 billion years ago from interstellar gas and dust that was enriched in heavier elements by the stars which were the successors to the first generation. Its Main Sequence lifetime is 10 billion years according to current stellar evolution models.

The lifetimes of smaller stars can be considerably longer, thanks to the period-luminosity relationship. The Main Sequence luminosity of a star is proportional to

NGC 6946 lies 25 million light years away, straddling the boundary between Cepheus and Cygnus. It was discovered by William Herschel in 1798. It is classified as an active starburst galaxy due to its very high rate of star formation. (NASA, ESA, STScI, R. Gendler, and the Subaru Telescope (NAOJ))

the fourth power of its mass, so a 0.1 solar mass star has a luminosity $1/10,000^{th}$ of the Sun. Thus, whilst it contains only $1/10^{th}$ of the hydrogen compared to the Sun, it burns it at a much lower rate, and ekes out its supply for a thousand times longer than the Sun. The lowest-mass red dwarfs could shine for a hundred trillion (10^{14}) years.

It has been estimated that 80% of all stars ever formed have yet to experience any significant stellar evolution. And there is still plenty of primordial hydrogen to form new stars. If you look at any photograph of a spiral galaxy, you can see star formation in progress. The spiral arms are delineated by clusters of young blue stars, many of them just a few million years old.

Eventually, however, the hydrogen will run out. Estimates of when this will happen range from one to a hundred trillion (10^{12} to 10^{14}) years, depending on how effectively the hydrogen from older stars is recycled back into interstellar space to become the raw material for new stars. The upper end of this range is also the Main Sequence lifetime of the lowest-mass stars, so we may conclude that the Age of Stars will end in a few hundred trillion (10^{14}) years.

To put this into perspective, if we liken the Age of Stars to a human lifetime of roughly 30,000 days, the Universe is barely a few days old.

The Second Age of Stars
Brown dwarfs are "failed" stars – bigger than gas giant planets, but not massive enough for nuclear fusion to occur at their cores. If two brown dwarfs were to collide and merge, the resulting object might be massive enough to become a very low mass star. It is estimated that the galaxy may contain as many brown dwarfs as normal stars – roughly a hundred billion. But space is vast, so collisions between brown dwarfs are extremely rare. The average time between such collisions in a galaxy like own is on the order of a hundred billion (10^{11}) years. After the end of the first Age of Stars, collisions between brown dwarfs could produce enough new stars to ensure that there are still around fifty hydrogen-burning stars shining in our galaxy at any moment.

The End of Baryonic Matter
Atoms are made from protons, neutrons and electrons. A free neutron (one that is not bound inside an atom) is unstable. It will decay into a proton, an electron and an anti-neutrino with a half life of around ten minutes.

Some Grand Unified Theories of sub-atomic physics predict that protons are also unstable, and will decay into a positron (an anti-electron) and a neutral pion. The half-life for this process is extremely long, and it has not been measured experimentally, but it is certainly greater than 10^{32} years, and less than 10^{49} years.

If the proton is indeed unstable, this has dire consequences for all "normal" baryonic matter in the universe, which will eventually decay into photons on these timescales. This process is a potential energy source for white dwarfs, although a typical white dwarf would produce less than 400 watts by proton decay, the equivalent of a few domestic light bulbs, and an entire galaxy of such stars would have less than a trillionth ($1/10^{12}$) of the luminosity of the Sun.

The End of Black Holes

Cosmologists now agree that at the heart of every galaxy, there is a super-massive black hole. By observing the movement of stars close to the centre of our own galaxy, astronomers have confirmed the existence of a black hole with the mass of 4.3 million Suns. These black holes will outlive all of the stars in their galaxies, and since they are not made of "normal" matter, they are immune to proton decay. But even black holes do not live forever. The cosmologist Stephen Hawking demonstrated that every black hole eventually "evaporates" via a quantum mechanical process known as Hawking radiation. The more massive the black hole, the longer this takes. A black hole like the one at the centre of our galaxy will take 10^{83} years to evaporate, whilst a black hole containing the mass of an entire galaxy will evaporate in 10^{98} years.

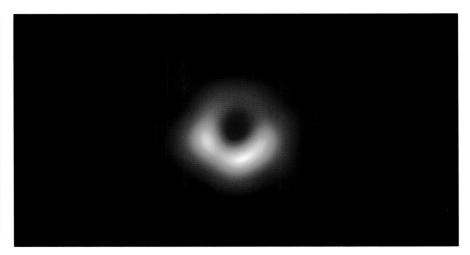

In 2019, an international network of radio telescopes was combined to create the Event Horizon Telescope. It was used to obtain this image of the 6.5 billion solar mass black hole at the centre of M87, outlined by emissions from hot gas circling the event horizon. (Event Horizon Telescope collaboration et al)

The Dark Era

At about 10^{100} years, the Universe will contain only electrons, positrons, neutrinos, and photons of colossal wavelengths. Beyond this point, our current knowledge of physics can say almost nothing. The Universe will have entered its Dark Era.

The End

Further Reading

Adams, D. *The Restaurant at the End of the Universe*. (Pan Books, 1980)

Caldeira, K, Kasting, J.F. (1992) 'The life span of the biosphere revisited.' *Nature* **360**, 721–723. (DOI: 10.1038/360721a0)

Adams, F.C., Laughlin, G. (1997) 'A dying universe: the long-term fate and evolution of astrophysical objects.' *Reviews of Modern Physics* **69**, 337–372 (DOI: 10.1103/RevModPhys.69.337)

Korycansky, D.G., Laughlin, G. Adams, F.C. (2001) 'Astronomical Engineering: A strategy for modifying planetary orbits.' *Astrophysics and Space Science* **275**, 349–366 (DOI: 10.1023/A:1002790227314)

Schröder, K.-P., Smith, R.C. (2008) 'Distant future of the Sun and Earth revisited.' *Monthly Notices of the Royal Astronomical Society* **386**, 155–163 (DOI: 10.1111/j.1365-2966.2008.13022.x)

The Evolution of "Multi-Pixel" Radio Telescopes

Rod Hine

In these days when almost everyone carries a smart phone with a sophisticated camera and probably watches high-definition TV, there can be few people who don't have some understanding of the concept of a pixel. Each pixel, or "picture element", represents a tiny point in the camera sensor capable of producing a signal representing the colour and intensity of the light from a spot in the image. In the TV screen, each pixel can emit light of specific colour and brightness, which, together with perhaps millions of other pixels, forms the image we see. The lens in our eye forms an image on the retina at the back of our eye which is formed of around a hundred million sensing elements, rods and cones, which likewise send information to the brain, from which we can perceive the visual world. In general terms, the more pixels the better.

The invention of the telescope around 1608 is usually attributed to Hans Lippershey in the Netherlands. He filed for a patent, although the use of lenses to enhance human vision was fairly widespread and the patent was not granted on the grounds that the device was already well-known. However, Lippershey benefited from the Dutch government by being given a contract to supply telescopes for official and military use. The combination of lenses in the telescope and the formidable image processing capabilities of the eye and brain can give amazing results, whether watching battles, birds, sports events or astronomical objects.

The average person now has such an instinctive understanding of the use of the telescope to form an image that when considering the term "radio telescope" he or she may reasonably ask "yes, but what can you see through it?" The answer is, disappointingly, not a lot! We must understand that the visible light that enters the eye has a wavelength between about 450 to 750 nano-metres. Around 2,000 wavelengths will fit into the space of a millimetre and as the rods and cones are tiny structures, they form a tightly packed array across the retina. By contrast the wavelengths used by radio astronomers range from a few metres to millimetres or so. However big the reflector, the received signals need to be converted into an electrical signal and the device that does that most efficiently, the dipole, has to have dimensions comparable to at least half a wavelength. So, most radio telescopes, even today, collect radio waves in a dish and focus them onto a single feed point and produce just one signal stream. The signal may have a wide range of wavelengths, corresponding to a wide spectrum of "colours" but it is essentially a single pixel.

To form a recognizable "image", therefore, the radio astronomer needs to perform extensive processing. In the early days of radio astronomy, the signals were filtered and recorded on long paper strips using chart recorders. By a tedious and painstaking process, the telescope could be scanned across the sky and an image reconstructed as a map of the intensity of the signals. Very early radio telescopes had little or no steerability, so the image was built up by the movement of the Earth spinning on its axis, a strip at a time. Later, the use of computers to scan steerable dishes across the sky and directly process the signals made the task a lot easier, but it still requires many hours of telescope time to produce a decent map, or image, of even a small portion of the radio sky.

The ultimate resolving power of any lens or mirror depends upon the diameter of the aperture relative to the wavelength. A typical modest optical telescope of 30 cm aperture should easily resolve two objects separated by just 0.4 arc-seconds in the sky. By contrast, even a 10 metre dish, operating at the 21cm H1 wavelength, can only resolve to about 1.5 degrees. Not only did the first radio maps of the sky take a lot of effort to construct, they were crude and blurry by comparison with optical images.

Jodrell Bank is now a World Heritage Site. (Anthony Holloway)

Early workers in this field devised various ingenious methods to overcome the limitations of their instruments in the quest for better resolution and speedier collection of data. In post-war UK, the two pioneering groups were at University of Manchester, under Bernard Lovell, and University of Cambridge, led by Martin Ryle. It proved impossible to work from Manchester itself, due to excessive levels of interference, and Lovell and his colleagues established themselves at Jodrell Bank in 1948. Working at first with ex-Army radar equipment they tackled the problem directly by building bigger dishes to improve resolution. After gaining experience with a 218 foot (66 metre) dish with limited steerability, Lovell conceived the ambitious idea of a fully steerable 250 foot (76 metre) dish. Finally completed in 1957 after incredible trials and tribulations,[1] the Mark 1, now known as the Lovell Radio Telescope, became an icon of the new technological age and in July 2019 was listed as a UNESCO World Heritage Site.

Also working at Jodrell Bank were astronomers Robert Hanbury-Brown and Henry Proctor Palmer, who, in 1954, devised a way to improve the resolution by using signals from two antennas. Their ingenious solution was called the rotating lobe interferometer[2] and used a complicated arrangement of motors, magslips and other electro-mechanical devices, doubtless salvaged from old radar equipment. Despite promising results, most of the effort at Jodrell Bank eventually went into completing the Mark 1 so the rotating lobe interferometer project just faded away. However, the germ of the idea, namely combining two or more signals using varying phase shifts, would later re-appear in a different guise.

Across at Cambridge, Ryle's group took a different approach, using the principles of interferometry and aperture synthesis to achieve better resolution by combining the signals from an array of a number of smaller dishes spread over a long distance. The Mullard Radio Astronomy Observatory (MRAO) was located at Lords Bridge, just west of Cambridge, and took advantage of a stretch of disused railway track to build an array of eight 13 metre dishes over a total length of about three miles. When fully deployed it was equivalent in resolving power to a dish of five kilometres diameter. However, achieving better resolution at longer wavelengths of the order of several metres was still a problem.

Another group at Cambridge, under Antony Hewish, built a huge array of wire dipole antennas in a large field adjacent to MRAO to make an instrument working at 85MHz (3.5 metres wavelength) with excellent resolution, high bandwidth and extreme sensitivity, but relying on the Earth's rotation to scan the sky. It was this

1. *The Story of Jodrell Bank*, Bernard Lovell Harper & Row, 1968.
2. *Out of the Zenith: Jodrell Bank 1957-1970*, Bernard Lovell, Oxford University Press, 1973.

Lords Bridge antennas at the Mullard Radio Astronomy Observatory. (Mike Lancaster)

instrument which led to the serendipitous discovery of pulsars in 1967. Jocelyn Bell, now Dame Jocelyn Bell Burnell, was the researcher who spotted some anomalous signals or "scruff" on the chart recordings. Her curiosity and patience in following up the initial observations paid off and after several more similar signals were observed, Hewish and Ryle went on to receive the Nobel Prize for Physics in 1974. Somewhat controversially, Jocelyn Bell was not included in the prize nomination. But the same "single pixel" limitation still meant a lot of tedious and time-consuming work to extract results from the data, and Hewish's instrument was later enlarged and improved in 1978 and continued in use for several more years. By the time of its final refurbishment in 1989 it employed the phasing techniques (see below) to generate 14 beams and was thus able to survey the whole northern sky in a day.

By the early 1960s, developments in microwave antenna design for radar applications had led to the concept of "phased array" antennas, echoing the work of Hanbury-Brown and Palmer. Using a linear array of small antennas, each fed with a signal whose phase was shifted slightly, it proved possible to effectively steer the beam by changing the phase shift. In 1961 Jesse Butler and Ralph Lowe published an article describing a simple but very clever method for producing the

From the left: Professor Stan Kurtz, Ernesto Andrade, Juan Luis Godoy and Pablo Villanueva at MEXART. Stan and Ernesto worked on the original design construction in around 2004, while Ernesto and Juan are developing the new digital receiver system, along with their assistant Pablo. (Rod Hine)

phase shift using only passive components, subsequently known as the "Butler Matrix".[3] This was used in the MEXART instrument built around 2004 to work at 138MHz (2 metre wavelength), operated by the National Autonomous University of Mexico (UNAM). The antenna consists of 64 rows of 64 dipoles, 4,096 separate dipoles altogether, and the 8x8 Butler Matrix gives 16 individual beams which can all work simultaneously. Recent dramatic improvements in digital signal processing (DSP) mean that it has now become worthwhile to replace the original passive Butler Matrix with a fully digital system for beam-forming. In its new incarnation, MEXART has much improved performance, more accurate calibration and sports no fewer than 64 beams.[4] However, MEXART still relies on the rotation of the

3. First described in the paper 'Beam-Forming Matrix Simplifies Design of Electronically Scanned Antennas', *Electronic Design*, 9, 170–173, 12 April 1961.
4. For more information see 'A New Digital Backend for the Mexican Array Radio Telescope' at **ieeexplore.ieee.org/document/8878959**.

ASKAP prototype phased array feed being installed at the Parkes Testbed Facility. (CSIRO Radio Astronomy Image Archive)

Earth to sweep the fan of beams across the sky. Like Hewish's array in Cambridge, its primary task is to investigate interplanetary scintillations due to the solar wind.

The latest step towards genuinely multiple-pixel radio telescopes has been taken with the construction of the Australian Square Kilometre Array Pathfinder or ASKAP, in Western Australia. Consisting of 36 dishes of 12 metres diameter, each dish is equipped with a phased-array antenna with 96 individual receiving sensors. Signals from the sensors are combined to form a bundle of 30 beams arranged as five rows of six. Using ASKAP, it is at last possible to produce a complete map of the radio sky at over a wide frequency range in a matter of hours and already ASKAP has provided new insights in radio astronomy.

The trend towards bigger dishes with exquisite precision of profile and therefore capable of working at smaller wavelengths, for example the Large Millimetre Telescope (LMT) in Mexico, may well see the development of even more exotic sensors than those fitted to ASKAP. It may then be possible to show the person in the street a "snapshot" with extreme detail taken with a radio telescope!

Mission to Mars
Countdown to Building a Brave New World
The Bare Necessities of Life

Martin Braddock

Introduction

The last article entitled 'Laying the Foundations' in *Yearbook of Astronomy 2022* illustrated the basic building materials present on the surface of the planet, how we can develop our understanding of their potential sourcing on Mars and characterise various terrestrial Martian regolith simulants until samples of authentic Martian regolith can be returned to Earth. In this article, we will now assume that a convoy of robotic 3D printers have been successfully deployed on Mars and that the construction of structures which will become our new homes is well underway.

Not withstanding the technical challenges of robotically building complex structures on the surface of another planet capable of providing the basics of shelter from the Martian environment, mission planners have to consider both physical and mental well-being of the prospective new arrivals. This starts with the journey in space. Launch windows are timed to coincide with the closest distance between Earth and Mars and are typically every two to three years where the interplanetary distance at opposition is between 58 and 92 million kilometres (Biswal and Annavarapu 2021). The actual journey follows a least energy Hohmann transfer orbit, which is an elliptical orbit used to transfer between the two circular planet orbits using the lowest possible amount of propellant. In the Hohmann transfer the spacecraft leaves the Earth at the transfer orbit perihelion and the transfer orbit aphelion coincides with the orbit of Mars. The overall distance is much further than the direct 'as the crow flies' distance. For example the Mars 2020 mission has travelled over 234 million kilometres from Earth to Mars.

It Will Not be Plain Sailing!

The timing of these transition windows is important as it minimises both the duration of the journey and therefore the amount of 'fuel' for both the spacecraft and occupants. Secondly, it sets the time constraints for mission duration and any potential return trip back to Earth. Returning a spacecraft from Mars to Earth, unmanned or otherwise is not currently possible from a logistical perspective, as

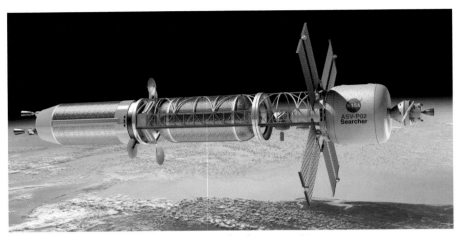

Getting there faster by advanced propulsion technologies! (NASA).

being able to carry enough fuel to take off from a planet which has a gravitational field of 0.38 g is prohibitive. This has implications as during an emergency situation – for example severe illness or injury, failure of the food or water supply chain or adverse Martian weather incidents – unlike astronauts in low Earth orbit or on the Moon, colonists of Mars will be unable to evacuate or be rescued. Having recognised the need to adhere to a pre-defined outbound transfer window, using current technology it takes between five and ten months to get to Mars; the recent Mars 2020 and Tianwen-1 missions took approximately seven months. This introduces a number of challenges. The first of these is the simple issue of food and water supply and the likely need to develop on board *in situ* resource utilisation technologies and capabilities, for example growing food on the journey and further optimisation of recycling, including that of human waste. Secondly, as the spacecraft moves further away from the Sun, the solar flux reduces which may be problematic for maintenance of vehicle power and an acceptable temperature controlled environment for the astronauts.

We may not be able to reduce the time between transfer windows, though mitigation plans should include shortening the journey time, possibly to as short as three months. The feasibility of using advanced propulsion technologies such as nuclear-thermal and nuclear-electric[1] is a concept which has been considered by NASA and other space agencies for many decades.

---

1. nasa.gov/directorates/spacetech/nuclear-propulsion-could-help-get-humans-to-mars-faster

Plans are being developed by the space agencies to develop rockets for interplanetary travel using nuclear fission and fusion, which become the principle propellant after the rocket has left the Earth's atmosphere, allaying safety concerns. This technology – which utilises energy released from nuclear reactions to heat liquid hydrogen to approximately 2,430°C, around eight times higher than the temperature of nuclear power-plant cores – generates twice the thrust of chemical rockets.

Preparing Body and Soul for the Journey

In 'It All Starts With a Journey' in *Yearbook of Astronomy 2021*, I briefly described the main physical effects of microgravity on the human body. Since then, findings have been published (Garrett-Bakelman et al 2019) from a study conducted on the International Space Station (ISS) called the NASA Twin Study.[2] This study compared multiple biochemical, molecular and physiological parameters between Scott Kelly who spent nearly a year on the ISS and his genetically identical twin Matthew who remained on Earth. Data reported, consistent with other studies, showed changes in gene expression, retained ability to mount an immune response, and changes in telomere length – the short sequences of DNA found at the end of our chromosomes – which returned to normal six months later. Other changes included chromosomal inversions, and a change in cognitive function. Most changes experienced during spaceflight returned to baseline, however, a small number of genes involved in the immune system and DNA repair did not return to baseline after Scott's return to Earth. This may be important to measure for longer term space missions as they may be biomarkers for astronaut health.

Recently NASA's Human Research Program Integrated Research Plan[3] has been updated to review the highest priority human health risks for a mission to Mars (Patel et al 2020). This evaluation has been conducted in the context of the space exposome which is defined as the total sum of spaceflight and lifetime exposures to hazards and how they relate to genetics and determine whole-body outcome. Current protection measures against space radiation are unlikely to provide sufficient protection for interplanetary missions of a number of years' duration, where the lifetime exposure limit is likely to be exceeded. The National Council on Radiation Protection and Measurements (NCRP) has issued guidelines on the recommended lifetime exposure to radiation which has been revised over the years (Bloshenko et al 2021) and differs by sex and age. Career radiation exposure limits vary by age and are lower for younger astronauts as they have a longer life span and may have

---

2. nasa.gov/twins-study
3. humanresearchroadmap.nasa.gov

a greater probability of developing subsequent health problems in later life. Career radiation exposure limits also vary by sex and are lower for female astronauts who are perceived to be at generally higher risk of radiation induced cancer. NASA's current space radiation health standard, based upon a risk model intended to set a limit of no more than a 3% additional lifetime risk of radiation exposure-induced death, recommends a lifetime exposure limit ranging from 180 millisieverts (mSv) for a 30 year old woman to 700 mSv for a 60 year old man. The new proposed standard,[4] based upon a 35 year old woman as being representative of the most vulnerable population, sets a single lifetime occupational limit of 600 mSv and applies to all ages and sexes. This is crucial to take into account as, in addition to an outward launch window, a return trip to Earth – should it be possible in future – will set the minimum time astronauts can spend on Mars, which includes timing to a reverse launch window. It raises some ethical questions asked and answered as part of the now defunct[5] Mars One programme and reinforces the need to provide adequate shielding. The MARSHA project at AI SpaceFactory[6] is considering the use of mixtures produced *in situ* of basalt fiber and renewable bioplastic in a 3D printing process as low molecular weight plastic is deemed effective in shielding radiation.

Astronaut psychology is equally, if not more important, as well-being, alertness and clear thinking are prerequisite to the success of any mission. This is a multi-faceted and lengthy topic though two areas are worthy of discussion. The first relates to the effect of intermittent or sustained artificial gravity on subjects taking part in a 60 day head down bed rest study (Basner et al 2021) at a declination of 6°, widely used as an analogue of microgravity on cognitive function. This study found a small but statistically reliable slowing of cognitive performance across a range of cognitive functions most consistently for sensorimotor speed, although accuracy was unaffected. The study also found no evidence for an effect of either continuous or intermittent artificial gravity on either cognitive performance or subjective responses and although the authors clearly state some limitations of their study, it further exemplifies the need for constant monitoring and vigilance of astronaut mental function. Secondly, there will inevitably be a psychological toll of being confined within a spacecraft with other astronauts and we will visit this in the future articles. However, two recent publications draw some interesting parallels. The first paper (White 2021), described in the editorial (de Vries 2021), proposes that isolation may *induce* creativity, possibly as a consequence of the Overview Effect,

4. National Academies of Sciences, Engineering and Medicine. Space Radiation and Astronaut Health: Managing and Communicating Cancer Risks, Washington DC: The National Academies Press. **doi.org/10.17226/26155**

5. **mars-one.com/faq/health-and-ethics/is-this-ethical**

6. **aispacefactory.com/marsha**

described as experiencing the Earth and the universe from a vantage point that very few human beings have had the privilege to do. The second publication (Chterev and Panero 2021) refers to creative problem solving in potential life-threatening situations and parallels across the space and theatre industries. Although an over-simplification, it would appear that the best way of maintaining tolerance with others in a confined environment is to have periods of constant challenge, interspersed with sufficient relaxation time and this fine balance will likely only be determined by trial, and hopefully not error.

What do You *Need* to Live?

The prospect of spending the rest of one's life on Mars with no immediate possibility of return to Earth is prerequisite for our new colonists, so the questions become what is essential for human survival and what is a requirement to maintain a quality of life sufficient to motivate colonists and attract immigrants for the future? Much can be learnt from the many analogue isolation studies which have been specifically designed to mimic conditions – as near as possible on Earth – to those which may be encountered on Mars. Selected candidate analogue missions, their duration, number of occupants and principle findings are shown in the Table and a more detailed account of analogue missions can be found elsewhere.[7]

Name	Duration	Location	Subjects	Measurements and Key Findings
HI-SEAS	4, 8 and 12 months in 2016	Hawaii Big Island, USA	6 people (3 men, 3 women)	Psychological and behavioural. Crew self-organization, group dynamics and living habits. Importance of personal space increased with mission duration
MDRS	14 day, 2013	Utah, USA	7 people (4 men, 3 women)	Psychological, behavioural and performance. Major fluctuations in total mood disturbance ameliorated by exercise, but not by music, horticulture or gaming
Mars-500	520 day between 2007 and 2011	Moscow, Russia	6 men (3 Russian, 1 Italian, 1 French and 1 Chinese)	Physiological and psychological. 4/6 members reported sleep disorders, avoided exercise, hid from others and had a disrupted circadian rhythm
FMARS	July 2009	Devon Island, Canada	6 people (3 men, 3 women)	Psychological, behavioural and performance. Clear need to appoint a Mission Commander with strong leadership skills

Table showing selected Earth analogues simulating conditions on Mars.
HI-SEAS = Hawaii Space Exploration Analog and Simulation; MDRS = Mars Desert Research Station; FMARS = Flashline Mars Arctic Research Station (M. Braddock).

Practicing for Mars.
The Hawaii Space
Exploration and
Analog Facility.
(Michaela Musilova /
HI-SEAS)

The reader is recommended to a recent publication which describes in depth three autonomous long duration missions at the Hawaii Space Exploration and Analog Facility (Heinicke 2021). The 110 square metre dome facility is 2,500 metres above sea level, located in the Mauna Loa mountain range on Hawaii Island, and has hosted five successful long-duration (4 to 12 month) NASA Mars simulation missions. The studies provide a fascinating insight to human behaviour, ergonomics and organizational dynamics which are essential to understand in order to provide a harmonious environment conducive to positive spirit by ensuring equality of contribution and fair distribution of tasks with ample opportunity for rotation of jobs within the community.

What Do You *Want* to Live?

It is clear that any Earth-bound vision of utopia will not be replicated on Mars though key questions for colonists are how do we define utopia and what does it mean at the level of the individual and could it be provided by automated assistance?

A recent United Kingdom-based crowd-sourcing study illustrated both physiological and psychological obstacles which should be overcome to colonize Mars (Braddock et al 2020). In this study, socio-technical systems models which included cognitive work analysis methodology were employed to build a nine step abstraction hierarchy supporting a work domain analysis. The top-ranking sub-category from the 387 responders who provided input was the provision of an adequate food and water supply chain. The second ranking criterion related to mental health and provision of a supportive environment. Today, terrestrial technology is already developed or under development which will be able to help. We already have the Amazon Alexa, Apple's Siri, Google Assistant, and Samsung's

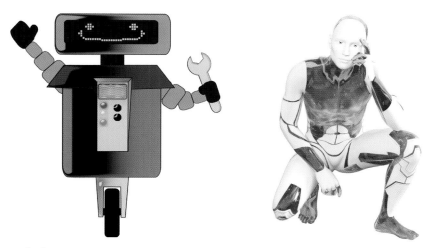

Hello, how are you today and how can we help? (PIXABAY)

Bixby platforms widely used on hundreds of millions of devices worldwide. Although the image of a free-wheeling robot shown below (left) appears cartoon-like and the image below (right) very futuristic, some technological developments beyond voice recognition have already been made and prototyped.

Facial recognition software is being developed by British firm Jaguar Land Rover to sense the stress levels of drivers and modulate the cabin conditions appropriately.[8] Other manufacturers are also exploring ways to monitor and correct our attention to driving[9] and a study has been conducted using emotion recognition software for the calibration of autonomous vehicles (Sini et al 2020). These examples serve to illustrate the potential for application of terrestrial technological virtual assistants to help assess, understand and perhaps stimulate the well-being of colonists going about their daily lives.

The Covid-19 pandemic turned the world upside down in 2020 and 2021 and human behaviour has shown remarkable adaptation for those able to access and work with science and technology. The next article in this series will address findings from the astrobiological and geological missions of Perseverance and the Zhurong rovers and serve as a primer for what we might find in the geology on Mars which helps us plan for future growth.

---

8. eandt.theiet.org/content/articles/2019/07/ai-mood-detection-in-cars-could-de-stress-driving
9. theverge.com/2018/4/13/17233614/renovo-affectiva-self-driving-car-driver-monitoring-emotion

References

Basner, M., Dinges, D.F., Howard, K., Moore, T.M., Gur, R.C., Mühl, C., Stahn, A.C. (2021). 'Continuous and Intermittent Artificial Gravity as a Countermeasure to the Cognitive Effects of 60 Days of Head-Down Tilt Bed Rest'. *Frontiers in Physiology*, 12. **doi.org/10.3389/fphys.2021.643854**

Biswal, M.K., Annavarapu, R.N. (2021). 'Interplanetary Challenges Encountered by the Crew During their Interplanetary Transit from Earth to Mars'. **arXiv:2101.04723**

Bloshenko, A.D., Robinson, J.M., Colon, R.A., Anchordoqui, L.A. (2021). 'Health threat from cosmic radiation during manned missions to Mars'. **arXiv:2012.09604v2**

Braddock, M., Wilhelm, C.P., Romain, A., Bale, L., Szocik, K. (2020). 'Application of socio-technical systems models to Martian colonisation and society build'. *Theoretical Issues in Ergonomic Science*, 21, 131–152. **tandfonline.com/doi/abs/10.1080/1463922X.2019.1658242?journalCode=ttie20**

Chterev, K., Panero, M.E. (2021). 'Exploring Similarities Across the Space and Theater Industries'. *Frontiers in Physiology*, 11. **doi.org/10.3389/fpsyg.2020.574878**

de Vries, H. (2021). 'Editorial: Creative Performance in Extreme Human Environments: Astronauts and Space'. *Frontiers in Physiology*, 12. **doi.org/10.3389/fpsyg.2021.709676**

Garrett-Bakelman, F.E. and 81 others. (2019). 'The NASA Twins Study: A multidimensional analysis of a year-long human spaceflight'. *Science*, 364. **science.sciencemag.org/content/364/6436/eaau8650**

Heinicke, C., Poulet, L., Dunn, J., Meier, A.J. (2021). 'Crew self-organization and group-living habits during three autonomous, long-duration Mars analog missions'. *Acta Astronautica*, 182, 160–178. **researchgate.net/publication/348983338**

Patel, Z.S., Brunstetter, T.J., Tarver, W.J., Whitmire, A.M., Zwart, S.R., Smith, S.M., Huff, J.L. (2020). 'Red risks for a journey to the red planet: The highest priority human health risks for a mission to Mars'. *npj Microgravity*, 6, Article 33. **doi.org/10.1038/s41526-020-00124-6**

Sini, J., Marceddu, A.C., Violante, M. (2020). 'Automatic Emotion Recognition for the Calibration of Autonomous Driving Functions'. *Electronics*, 9, 518. **doi.org/10.3390/electronics9030518**

White, W.F. (2021). 'The Overview Effect and Creative Performance in Extreme Human Environments'. *Frontiers in Physiology*, 12. **frontiersin.org/articles/10.3389/fpsyg.2021.584573**

The Ability to Believe
The Bizarre Worlds of Astronomical Antireality

Neil Haggath

"Faith is the ability to believe what you know not to be true."
Mark Twain

The town of Zion, Illinois, was founded in 1901, as a home for a group of religious zealots called the Zion Tabernacle of the Christian Catholic Apostolic Church. No other church was allowed to be established in the town, and any visiting preacher of any mainstream denomination was quickly expelled. Those of the faith were subject to all the usual kinds of prohibitions which one would expect of such a sect – but there was one particularly bizarre one… no shop in the town was allowed to sell globes! That was because the sect's founder, a self-proclaimed "faith healer" named John Alexander Dowie (1847–1907), insisted that his followers believe that the Earth was flat!

John Alexander Dowie. (John McCue)

You will probably not be surprised to learn that Dowie was in fact a spectacularly successful fraudster! He literally owned the entire town; all the residents had to rent their properties from him, and deposit their savings in his bank. While commanding his "disciples" to observe a life of "humility and self-denial", he and his family enjoyed extravagant wealth.

But what of his Flat Earth belief? Apparently, that arose from an insistence that every word of the Bible must be taken literally, and a refusal to accept that phrases such as "to the four corners of the Earth" are merely figures of speech. (So the flat Earth is in fact square!)

The classical Greeks deduced that the Earth is spherical around 2,500 years ago, by observing the shape of its shadow during lunar eclipses, and in 240 BC,

Eratosthenes determined its size with remarkable accuracy. The popular notion that Columbus "wanted to prove that the Earth was round", and that people were "afraid he would fall off the edge", is complete nonsense; he was in fact trying to establish a new trade route to India and the Far East. No-one with even a modicum of intelligence thought the Earth was flat by 1492!

So how is it that even today, there are still people in the world who believe exactly that? The simple answer is that their beliefs have nothing to do with science or reason, and everything to do with dogmatic religion. But there is more to it.

A friend of mine, the late Ray Worthy, wrote an article for our astronomical society's magazine, titled 'The Town That Declared the Earth was Flat', in

Samuel Birley Rowbotham's Flat Earth map, from *Zetetic Astronomy*, 1865. (Samuel Birley Rowbotham)

which he examined the above and other stories of Flat Earth believers. Ray was convinced that modern day Flat-Earthism can be traced to the "teachings" of one Samuel Birley Rowbotham (1816–84), who founded a forerunner of what is now the Flat Earth Society. But he too was nothing more than a charlatan, who invented his "belief" for the sole purpose of parting the gullible from their money!

Rowbotham was, in Ray's words, "a charismatic snake oil salesman" – almost literally, as he had indeed begun his career selling bogus "medicine" – who toured Britain giving public lectures, in which he presented seemingly plausible arguments for a flat Earth, and challenged people to debate him and try to prove him wrong. They paid sixpence a time for the privilege, and Rowbotham laughed all the way to the bank.

Alfred Russel Wallace c1895, image first published in *Borderland* magazine, April 1896. (Wikimedia Commons / London Stereoscopic and Photographic Company / Borderland Magazine)

Yet some people were completely taken in by him! In 1870, one of his "disciples", John Hampden, issued a wager in a reputable magazine, whereby he would pay a substantial sum to anyone who could demonstrate by experiment that the Earth is round. The challenge was taken up by no less than Alfred Russel Wallace (1823–1913) – the man who had proposed the Theory of Evolution by Natural Selection independently of Charles Darwin – who proceeded to do exactly that, to the satisfaction of the magazine's Editor, who was appointed as judge. You can read the details of the demonstration in Ray's article.[1]

But Hampden refused to accept defeat, and refused to look at the evidence. For years, he publicly accused Wallace of cheating him; until the day he died, he refused to accept that he had been wrong!

This was an extreme example of the behaviour typical of religious zealots, who stubbornly adhere to beliefs which defy all logic and reason, no matter what.

---

1. Ray Worthy, 'The Town That Declared the Earth was Flat', *Transit*, January 2015. **cadas-astro.org.uk/transitmag/Transit0115.pdf**

Today, of course, we have the Young Earth Creationists, who deny evolution – and therefore, effectively, the whole of modern science – and assert that the Earth was created 6,000 years ago, as "calculated" by the seventeenth century Archbishop Ussher. Again, their beliefs are based on an insistence that the Bible is "the inerrant word of God", and everything in it is literally true.

Allow me to clarify a point here. While I am an atheist and proud of it, I am *not* saying that every Christian, or everyone who believes in God, is a fool. I *am* saying that anyone who disputes the indisputable, and believes things which are completely and *provably* false, is deluded or misguided! There's a chasm of difference between believing in God and believing that the Bible is all true! The vast majority of Christians today, and all mainstream churches, readily acknowledge that the Bible is not, and was never meant to be, an accurate history book or scientific textbook, and accept such things as evolution and the age of the Earth.

In most of the western world, creationists are a small – but often very vocal – minority – though they seem to be increasing in numbers! The latter fact, to me, is a clear indication of the decline in education standards, and especially science education. Despite the fantastic potential of the internet as an educational tool, it seems to be used far more to propagate *mis*information and senseless drivel; the rise in popularity of "conspiracy theories" and other kinds of antireality stems from the attitude of, "It must be right, because I read it on the internet"!

In my home city alone, I have encountered several creationists, who loudly harangue passers-by about their pre-Darwinian beliefs. I sometimes get into "debates" with them, just for the intellectual fun of it. Invariably, they try to ridicule evolution, without having the slightest comprehension of what it is they are ridiculing – which is a guaranteed way of making oneself look an utter idiot! (I suspect that most of their fellow Christians consider them an embarrassment.)

They are also fond of claiming that "evolution is only a theory" – which simply demonstrates their own ignorance of the scientific method, and what a theory *is*. They apparently think a theory "becomes a fact when it's proved". It does not! Strictly speaking, a theory can never be absolutely proven - but the Theory of Evolution by Natural Selection is supported by such an overwhelming amount of evidence that it can be considered proven beyond reasonable doubt, and as close to established fact as any theory ever *can* be!

In the United States, however, creationists now make up a substantial proportion of the population. Disturbingly, they include many in very high positions of authority and political power – up to and including a recent Vice-President! There have been many attempts to force schools to teach Biblical creationism, not only in religious studies, but in *science* classes, as a "valid alternative theory to evolution"

– it is of course no such thing! Such efforts are, thankfully, always doomed to failure. The country's constitution guarantees the separation of church and state, and forbids the state or government from promoting or supporting any single religion above others. Nevertheless, the state of Texas, in a frightening move of monumental stupidity, appointed a notoriously scientifically ignorant creationist as Chairman of its State Board of Education!

The American science populariser Bill Nye – who presents a TV series as "Bill Nye the Science Guy" – gave a public talk in a certain town in Texas. A substantial proportion of the audience booed him, and several walked out, because - and this truly defies belief - he told them that the Moon does not "shine" in its own right, but only by reflecting sunlight!

Apparently, those people take literally Genesis 1:16: "And God made two great lights; the greater light to rule the day, and the lesser light to rule the night; He made the stars also." Yes, *really*! Again, the Greeks deduced that the Moon only shines by reflected sunlight, two millennia ago.

So if those people believe that the Moon is a "light", then how do they explain its phases? And if it "rules the night", how come it's frequently seen in daylight, and *not* at night? Oh, but wait… an appalling proportion of adults apparently don't even *know* that the Moon is visible in daylight!

A few years ago in my home city, someone called a phone-in programme on the local radio station, expressing amazement because he could see the Moon in

A puzzling phenomenon – a daylight Moon for all to see! (Mary McIntyre)

daylight. The host was equally baffled, and asked if anyone could phone in, who could explain this puzzling phenomenon! So it seems that a significant proportion of adults actually believe the childish notion that the Moon "only comes out at night"!

Conversely, however, another astronomer friend told me that his daughter realised for herself, with no prompting, that she could see the Moon in daylight, at the age of five.

While I have never personally encountered a Flat-Earther, I *have* come across something equally absurd. Around ten years ago, I engaged in a truly mind-boggling "debate" in an online forum, with a fellow who is not only a creationist, but also a "Moon landings were faked" conspiracy theorist, *and* – wait for it… a *geocentrist!*

Yes, you *did* read that correctly! There are actually still people who believe, in the twenty-first century, that the Earth is the centre of the Universe, and everything else revolves around it! To put it another way, they reject the last four centuries of scientific discovery, the whole of Newtonian, let alone Einsteinian, physics, and effectively don't believe in gravity!

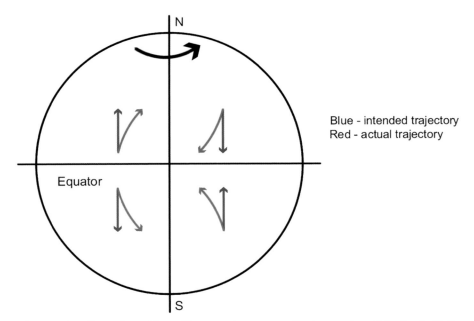

The Coriolis Effect: why artillery officers need to account for the rotation of the Earth. Shells follow curved trajectories with respect to the Earth's surface, as points at different latitudes are moving eastward at different linear velocities. (Neil Haggath / Garfield Blackmore)

The reason that this fellow believes that is, quite simply, because the Bible says so. Except it doesn't! I have two Christian friends, of the rare kind who have actually read the entire Bible; both tell me that it says no such thing. There are a few ambiguous statements, which one could perhaps choose to interpret that way, but nowhere does it explicitly say so.

Again, I "debated" this fellow, just for the sheer intellectual fun of demolishing his ridiculous arguments. Among his many delusions is the one that not only Apollo, but *every spaceflight in history*, was and is faked! He claims that practically the whole of modern science is a gigantic worldwide conspiracy, to falsely "prove Galileo right", and convince us all of the "false theory" of heliocentrism!

Needless to say, he made no attempt to say *why* such a conspiracy would exist! For what conceivable purpose does he imagine that the entire scientific community, worldwide, would conspire to convince everyone else of something which isn't true? And above all, what would be the point of *any* conspiracy, which by its very nature, would require a substantial proportion of the world population to be "in on it"?

But even that isn't the end of it. While he accepts that the Earth is spherical, he claims that it doesn't rotate. I countered that by saying, "In that case, not only is every scientist in the world part of your gigantic conspiracy, but every artillery and gunnery officer in every army and navy in the world must also be in on it!" That point went right over his head; he had no idea what I was talking about, and I had to explain the Coriolis Effect. I trust that it isn't lost on you, dear reader…

For all that we depend on science in our modern world, the deniers of reality are all around us!

The Astronomers' Stars
Life in the Fast Lane

Lynne Marie Stockman

For over 2,000 years, humans have been cataloguing the heavens, noting the positions and brightnesses of the stars. The accuracy improved as the instrumentation and observational techniques did. And as natural philosophers and scientists began comparing their observations with those of their predecessors, they noticed something odd: some of the fixed stars were not so fixed.

Giuseppe Piazzi (1746–1826) was born in Ponte in Veltellina, Italy. Little is known of his early life but he was a priest in the Theatine Order of the Catholic Church. He studied at a number of colleges of that order where he learned mathematics and astronomy, and later taught philosophy, mathematics and theology. He went to the academy at Palermo (now the University of Palermo) in 1780 to take up the chair of higher mathematics and obtained a grant from the viceroy of Sicily to construct an observatory. In preparation, Piazzi studied with Joseph Jérôme Lefrançois de Lalande in Paris and with the Reverend Dr Nevil Maskelyne, the Astronomer Royal, in England. Whilst in England he met Jesse Ramsden, a noted English scientific instrument maker. Ramsden made several instruments for the fledgling observatory at Palermo and in 1791, observations began, leading to the 1803 publication of the star catalogue *Præcipuarum Stellarum Inerrantium Positiones Mediæ Inuente Seculo XIX: ex Observationibus Habitis in Specula Panormitana ab anno 1792 ad annum 1802* with a second edition coming out in 1814. Piazzi is probably better remembered for his discovery of the first asteroid (now dwarf planet), Ceres, in 1801. The Académie des sciences awarded him the Prix Lalande twice, in 1804 and 1814. He died in Naples in 1826. (Fox 1911)[1]

61 Cygni is the Flamsteed designation of a binary system consisting of two K-type stars, one being fifth magnitude and the other sixth. Both stars exhibit solar-like activity cycles, and the magnetic field of 61 Cygni A reverses in phase with its chromospheric and coronal cycles, much like the Sun does (Boro Saikia et al 2016). 61 Cygni was included in the 1803 Palermo catalogue where Piazzi not

---

1. References are found at the end of the article.

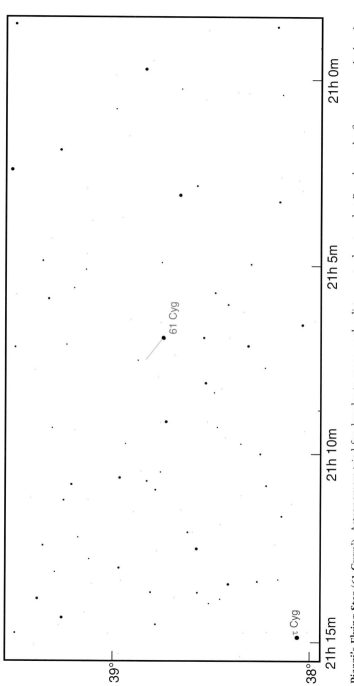

Piazzi's Flying Star (61 Cygni). Astronomers tried for decades to measure the distances to the stars, but Bessel was the first to succeed when he measured the parallax of 61 Cygni in 1838. His result is within 10% of the modern value. This finder chart is a 3.4° by 1.6° field with 61 Cygni at the centre *at its location in 2000*. The line shows its motion in one century, from 2000 to 2100. Stars are shown to magnitude +10.0 with the brightest being the δ Scuti variable star τ Cygni (+3.7). Coordinates are referred to the epoch J2000. (David Harper)

only published his coordinates for the star but also the difference in position from Flamsteed's earlier observations (Piazzi 1803). The difference was startling, and handwritten notes in Piazzi's own copy of the catalogue showed that he was well aware of the discrepancy (Foderà-Serio 1990). Three years later Piazzi published another smaller catalogue of stars with their proper motions. On page 30 in this volume he compared his observed position of 61 Cygni with the observations of English astronomers John Flamsteed and James Bradley and on page 10 stated (Piazzi 1806):

> Sebbene le stelle tutte abbiano forse movimenti proprj, nientedimeno, per quanto io possa giudicarne, noi non possiamo contarne che una sessantina, delle quali non sia permesso di dubitare. I maggiori sono nelle costellazioni del Cigno, di Cassiopea, e dell' Eridano (a).
>
> (a) Comunque si combinino sì fatti movimenti non veggo in quale maniera si possano spiegare per quello del nostro sistema. Quindi, se sono essi traslazioni reali delle stelle a diversi punti dello spazio, le distanze delle stelle medesime da noi debbono avere un rapporto finito col diametro dell' orbita terrestre, rapporto, che in alcune almeno, come sono μ Cassiopea, 61 del Cigno, D dell' Eridano, non può non essere molto sensibile. Osservate per tanto queste stelle ne' tempi delle loro paralassi, massima e minima, dovrebbero presentare delle differenze, atte a togliere ogni dubbio.

> Although the stars may all have proper motions, nevertheless, as far as I can judge, we can count only about sixty of which there is no doubt. The major ones are in the constellations of Cygnus, Cassiopeia, and Eridanus (a).
>
> (a) No matter how these movements are combined, I don't see how they can be explained in our system. Thus, if they are real translations of the stars at different points in space, the distances of the stars themselves must have a finite relationship with the diameter of the earth's orbit, which in some at least, such as μ Cassiopeiae, 61 Cygni and D Eridani, cannot be very sensitive. Therefore, we should observe these stars in their parallaxes, maximum and minimum, should they present some differences, to remove any doubt.

Piazzi's Flying Star, as it later became known, was independently investigated by German astronomer **Friedrich Wilhelm Bessel** (1784–1846). Born in Minden, Westphalia, his early education was not a success and he was apprenticed to a mercantile firm in Bremen in 1799. During this time he studied mathematics and astronomy on his own, and in 1806 he became an assistant astronomer at a private observatory near Bremen. Whilst there, Bessel calculated the orbit of Halley's Comet and made numerous observations of comets and planets. Three years later he was appointed the director of the new observatory in Königsberg where he remained for the rest of his career. His work at Königsberg included determining the positions of thousands of stars (assisted by F.W.A. Argelander) and developing a fundamental reference frame. Bessel's work won him the Gold Medal of the Royal Astronomical Society in 1829 and again in 1841, and he was twice awarded the Prix Lalande (1811, 1816) by the Académie des sciences. He died from cancer in 1846 and was buried near his beloved observatory (Fricke 1985).

In 1838, Bessel wrote to English astronomer Sir John Herschel:

> After so many unsuccessful attempts to determine the parallax of a fixed star, I thought it worth while to try what might be accomplished by means of the accuracy which my great Fraunhofer Heliometer gives to the observations. I undertook to make this investigation upon the star 61 *Cygni*, which, by reason of its great proper motion, is perhaps the best of all; which affords the advantage of being a double star, and on that account may be observed with greater accuracy; and which is so near the pole that, with the exception of a small part of the year, it can always be observed at night at a sufficient distance from the horizon.

In the letter, Bessel presented his measurements of each component of the star and explained how he calculated the parallax, ending with value of $0''.3136$. This translates into a distance of 3.189 pc. Modern values from the *Gaia* data set suggest a distance of 3.497 pc for 61 Cygni A and a slightly closer distance of 3.495 pc for 61 Cygni B. With his careful observations, Bessel became the first astronomer to directly measure the distance to a star outside of the solar system. For this reason, 61 Cygni is sometimes called ***Bessel's Star***.

Friedrich Wilhelm August Argelander (1799–1875) was born in Memel in the Kingdom of Prussia (now Klaipėda in Lithuania) and studied under Bessel at the University of Königsberg. He earned his PhD in 1822 and later that year wrote a treatise on the orbit of the great comet of 1811 (now formally designated C/1811 F1) which was received with considerable acclaim throughout Europe. On Bessel's recommendation, Argelander was appointed head of the Finnish Observatory at Åbo (Turku) in 1823. During his time there, Argelander made a special study of stars suspected of having large proper motions. Fire ravaged the town and university of Åbo/Turku in 1827 and although the observatory wasn't damaged, it was decided to move it (and the university) to Helsingfors. The observatory finally opened in 1835. However, Argelander moved to Bonn two years later to take up directorship of the new observatory that was to be built there. It was ten years before the observatory was completed and in the meantime, Argelander used what instruments he had to continue his observations. Bessel had earlier hit upon the idea of cataloguing all the stars down to ninth magnitude but nothing had come of it. With the completion of the Bonn Observatory, Argelander now had a chance to make that dream a reality and in 1852, began the work that would eventually result in the influential star catalogue *Bonner Durchmusterung*. He was the recipient of the Gold Medal of the Royal Astronomical Society in 1863 and was a member of numerous European and North American scientific societies (Anonymous 1876).

Argelander's Star was first catalogued by Stephen Groombridge and appeared in his 1838 *Catalogue of Circumpolar Stars* as star number 1830. Argelander recognised that it had a high proper motion in 1841, commenting (Argelander 1842):

Im vorigen Jahre habe ich zufällig einen merkwürdigen Stern gefunden, einen Stern, dessen eigene Bewegung alle bekannten bedeutend übersteigt, indem sie 7″ im Bogen des gröfsten Kreises jährlich beträgt. Es ist dies ein Stern…Nr. 1830 in *Grombridges* Catalog von Circumpolarsternen.

Last year, by chance, I found a strange star, a star whose proper motion greatly outshines all the known ones, being 7″ of arc a year. It is the star…No. 1830 in Groombridge's Catalogue of Circumpolar Stars.

Recent results from the *Gaia* astrometry mission confirm a proper motion of 7″.06 per year, the third largest yet catalogued. Like the Sun, Argelander's Star is a G-type dwarf, but that is where the similarity ends. It is a sixth-magnitude Population II star originating in the halo of our galaxy, making it much older than the Sun and somewhat more metal-deficient. In 1939, the star experienced a superflare, a very strong explosion thousands of times more energetic than a typical solar flare (Schaefer et al 2000). Another possible flare was recorded in the 1960s and Argelander's Star has since received the variable star designation CF Ursae Majoris.

Argelander's Second Star first appeared in a star catalogue as number 21185 in Lalande's 1801 *Histoire Céleste Française*. At magnitude +7.5, it is the brightest red dwarf star in the northern celestial hemisphere. Like Argelander's Star, it is located in the constellation of Ursa Major. Argelander noticed its high proper motion in 1857 when he wrote:

Vor einigen Wochen hat unsere Durchmusterung wieder einen Stern mit sehr starker Bewegung erkennen lassen. Es ist dies der Stern L.L. No. 21185....

A few weeks ago, our survey again showed a star with very strong movement. It is the star L.L. No. 21185....

Argelander's Second Star has a smaller annual proper motion (4″.80) than Argelander's first star but in the following year, German astronomer Friedrich August Theodor Winnecke measured its parallax as 0″.511 which is equivalent to a distance of 1.96 pc (Winnecke 1858). (The modern parallax value from Hipparcos is 0″.39264 which converts to a more distant 2.55 pc from the Sun.) Until the discovery of Barnard's Star in 1916, Argelander's Second Star was the nearest known single star to the Sun. Unseen low-mass companions for this faint star have been claimed since 1944, but more recent observations seem to confirm a super-Earth with a 13-day period revolving around the star (Stock et al 2020).

Jacobus Cornelius Kapteyn (1851–1922) was born in the Netherlands and was educated in Utrecht. He spent two years at Leiden Observatory and was appointed Professor of Astronomy at the University of Groningen in 1878 where he remained until retirement. Beginning in 1885 he started a long collaboration with Sir David Gill which eventually led to the publication of the star catalogue *Cape Photographic Durchmusterung*, a southern hemisphere counterpart to Argelander's earlier *Bonner Durchmusterung*. Kapteyn was interested in the distribution of stars in space, analysing their various properties using statistical methods. His work on stellar proper motions would later provide evidence of galactic rotation. He was awarded the Gold Medal of the Royal Astronomical Society in 1902 and was a research assistant at Mount Wilson Observatory in California which he visited on occasion. Kapteyn died in 1922, a year after he retired (Anonymous 1922).

The variable star VZ Pictoris is often called ***Kapteyn's Star***. This ninth-magnitude red sub-dwarf is a member of the galactic halo population and has an unusually high proper motion. Kapteyn drew attention to this star in 1897:

Stern mit grösster bislang bekannter Eigenbewegung
Der Stern Cordoba Zone Catalogue 5ʰ.243 hat eine Eigenbewegung von 8″.7 im grössten Kreise, wie dies aus folgenden Beobachtungen hervorgeht... welche alle gut stimmen zu einer Eigenbewegung von +0ˢ.621 in gerader Aufsteigung, und von −5″.70 in Declination.
Innes findet den Stern orange-gelb.

ω Centauri has been catalogued as a star since the time of Ptolemy's *Almagest* with Johann Bayer giving it its present designation in 1603. It was later discovered to be a non-stellar object and was eventually classified as a globular cluster. However, this object may actually be the remnant of a dwarf galaxy that collided with the Milky Way long ago. Kapteyn's Star is thought to have originated here. This false-colour image of ω Centauri combines the infrared observations of the Spitzer Space Telescope with visible light data from the National Science Foundation's Blanco 4-metre telescope at the Cerro Tololo Inter-American Observatory in Chile. (NASA/JPL-Caltech/NOAO/AURA/NSF)

Star with largest known proper motion
The star Cordoba Zone Catalogue 5^h.243 has a proper motion of $8''.7$ of arc, as can be seen from the following observations…which all agree well with a proper motion of $+0^s$.621 in right ascension and of $-5''.70$ in declination. Innes finds the star orange-yellow.

These values are in good agreement with recent *Gaia* measurements of $\mu_\alpha \cos \delta$ = 6″.491474 and μ_δ = −5″.709218 per year which makes for a total annual proper motion of 8″.64. Only Barnard's Star scoots across the sky more quickly. Kapteyn's Star also has a high radial velocity and orbits the galaxy in retrograde. It's likely that it originated in the globular cluster ω Centauri which itself may be the remnant of a dwarf galaxy that collided with the Milky Way long ago (Kotoneva et al 2005). Our near neighbour Kapteyn's Star is definitely from out of town and may have brought a planet or two with it (Anglada-Escudé et al 2014).

Edward Emerson Barnard (1857–1923) was born in Nashville, Tennessee, in the United States, his father having already died. He grew up in poverty, receiving little formal education, and was apprenticed to a photography studio at the age of nine. He taught himself the basics of astronomy and acquired his first small telescope in 1877, using it to observe the planets and to discover several comets. These early discoveries brought him to the attention of the Chancellor of the newly-established Vanderbilt University, and Barnard was put in charge of the university's observatory in 1883. He enrolled at the university at the same time although sources disagree as to whether his degree was honorary. His photographic skills, learned from a young age, stood him in good stead and eventually took him to Lick Observatory in 1887 where he developed techniques for photographing comets, nebulae and the Milky Way. He also discovered a satellite of Jupiter, the first since Galileo's telescopic investigation of the gas giant in the early seventeenth century. Barnard was appointed professor of astronomy to the University of Chicago in 1895 where he had access to Yerkes Observatory. He remained there for the rest of his life. He published over 800 articles covering a wide variety of subjects, was a pioneer in astrophotography, and produced the *Barnard Catalogue of Dark Markings in the Sky* in 1919. He was a member of a number of learned societies, and the recipient of many awards and prizes for his observational work, including the 1892 Prix Lalande from the Académie des sciences and the Gold Medal of the Royal Astronomical Society in 1897. He was an associate editor of *The Astronomical Journal* from 1914. He died in his home near Yerkes Observatory in 1923 (Anonymous 1923).

Barnard announced his discovery of the star that was to bear his name in 1916:

I have found on my photographs a small star of the 11th (photographic) magnitude which has a large proper-motion of about 10″.3 a year in a direction almost due north. The star is yellowish…A comparison with a photograph taken with the 10-inch Bruce telescope on June 11, 1904, gives the annual motion 10″.36. At present this is the largest known proper-motion of a star.

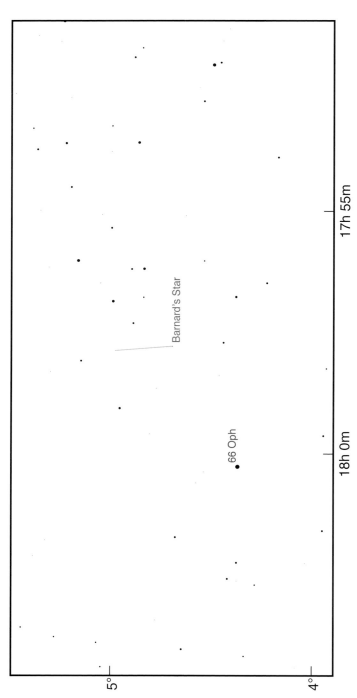

Barnard's Star exhibits the highest proper motion yet measured in any star and is the fourth-nearest star to the Sun. The International Astronomical Union officially recognised the name "Barnard's Star" in 2017. This finder chart is a 3.4° by 1.6° field with Barnard's Star at the centre *at its location in 2000.* The line shows its motion in one century, from 2000 to 2100. Stars are shown to magnitude +10.0 with the brightest being the blue Be star 66 Ophiuchi (+4.6). Coordinates are referred to the epoch J2000. (David Harper)

Barnard's Star is a low-mass red dwarf star which also carries the variable star designation V2500 Ophiuchi. Published results from *Gaia* yield an annual proper motion of 10″.39 and a distance of 1.83 pc. This makes Barnard's Star the closest single star to the Sun and the fourth closest after the trinary system of α Centauri. And it still holds the record for highest proper motion. However, Barnard's Star may not be alone. In 2018 an international team of astronomers announced the possible detection of a planet in orbit around the star. The planet-candidate is around three times the mass of the Earth with an orbital period of 233 days (Ribas et al 2018).

Acknowledgements

This research has made use of NASA's *Astrophysics Data System Bibliographic Services*, operated at the Harvard-Smithsonian Center for Astrophysics, Cambridge, Massachusetts, USA, and the SIMBAD astronomical database, operated at CDS, University of Strasbourg, France. The author would like to thank Dr David Harper for his enthusiastic encouragement, helpful comments and gentle introduction to the mysteries of astrometry. The author would also like to thank Dr Cinzia Malangone and Ms Dana Stockman for their assistance with the Italian translation.

References

[Anonymous] 1876, 'Obituary: Friedrich Wilhelm August Argelander', *Monthly Notices of the Royal Astronomical Society*, **36**, 151–155.

[Anonymous] 1922, 'Obituary: Jacobus Cornelius Kapteyn', *The Observatory*, **45**, 261–265.

[Anonymous] 1923, 'Obituary: Edward Emerson Barnard', *The Astronomical Journal*, **35** (4), 25–26.

Anglada-Escudé, G., Arriagada, P., Tuomi, M., and 21 others. 2014, 'Two planets around Kapteyn's star: a cold and a temperate super-Earth orbiting the nearest halo red dwarf', *Monthly Notices of the Royal Astronomical Society: Letters*, **443**, L89–L93. **doi.org/10.1093/mnrasl/slu076**

Argelander, F.W.A. 1842, 'Schreiben des Herrn Professors Argelander, Directors der Sternwarte in Bonn, an den Herausgeber', *Astronomische Nachrichten*, **19**, 393–396.

Argelander, F.W.A. 1857, 'Schreiben des Herrn Professors Argelander, Directors der Sternwarte in Bonn, an den Herausgeber', *Astronomische Nachrichten*, **46**, 273.

Barnard, E.E. 1916, 'A Small Star with Large Proper-Motion', *The Astronomical Journal*, **29**, 181–183.

Bessel, F.W. 1838, 'A letter from Professor Bessel to Sir J. Herschel, Bart., dated Konigsberg, Oct. 23, 1838', *Monthly Notices of the Royal Astronomical Society*, **4**, 152–161.

Boro Saikia, S., Jeffers, S.V., Morin, J., and 9 others. 2016, 'A solar-like magnetic cycle on the mature K-dwarf 61 Cygni A (HD 201091), *Astronomy & Astrophysics*, **594**, A29. **doi.org/10.1051/0004-6361/201628262**

Foderà-Serio, Giorgia. 1990, 'Giuseppe Piazzi and the Discovery of the Proper Motion of 61-CYGNI', *Journal for the History of Astronomy*, **21**, 276–281.

Fox, William. 1911, 'Guiseppe Piazzi' in *The Catholic Encyclopedia* (New York: Robert Appleton Company, Volume 12).

Fricke, Walter. 1985, 'Friedrich Wilhelm Bessel (1784–1846)—In Honor of the 200th Anniversary of Bessel's Birth' in *Astronometric Binaries, Astrophysics and Space Science* (eds. Z. Kopal, J. Rahe) (Dordrecht: D. Reidel Publishing Company), **110**, 11–19.

Kapteyn, J.C. 1897, 'Stern mit grösster bislang bekannter Eigenbewegung', *Astronomische Nachrichten*, **145**, 159.

Kotoneva, E., Innanen, K., Dawson, P.C., and 2 others. 2005, 'A study of Kapteyn's star', *Astronomy & Astrophysics*, **438**, 957–962. **doi.org/10.1051/0004-6361:20042287**

Piazzi, Giuseppe. 1803, *Præcipuarum Stellarum Inerrantium Positiones Mediæ Inuente Seculo XIX: ex Observationibus Habitis in Specula Panormitana ab anno 1792 ad annum 1802.*

Piazzi, Giuseppe. 1806, *Libro sesto del Reale Osservatorio di Palermo.*

Ribas, I., Tuomi, M., Reiners, A., and 60 others. 2018, 'A candidate super-Earth planet orbiting near the snow line of Barnard's star', *Nature*, **563**, 365–368. **doi.org/10.1038/s41586-018-0677-y**

Schaefer, Bradley E., King, Jeremy R., Deliyannis, Constantine P. 2000, 'Superflares on Ordinary Solar-Type Stars', *The Astrophysical Journal*, **529**, 1026–1030. **doi.org/10.1086/308325**

Stock, S., Nagel, E., Kemmer, J., and 37 others. 2020, 'The CARMENES search for exoplanets around M dwarfs. Three temperate-to-warm super-Earths', *Astronomy & Astrophysics*, **643**, A112. **doi.org/10.1051/0004-6361/202038820**

Winnecke, A. 1858, 'Über die Parallaxe des zweiten Argelander'schen Sterns', *Astronomische Nachrichten*, **48**, 289–292.

Elijah Hinsdale Burritt
Geographer of the Heavens

Richard H. Sanderson

The first half of the nineteenth century was a time of burgeoning interest in astronomy in the United States. People eagerly watched the 1806 total solar eclipse, marvelled at bright comets and were dazzled by a rare outburst of the Leonid meteor storm. America's first astronomical observatories began probing the heavens. Elijah Hinsdale Burritt made an important contribution to this remarkable half century by writing one of the best-selling American astronomy books of its day.

Born in New Britain, Connecticut in 1794, Elijah Burritt was the eldest of ten siblings and the brother of peace activist Elihu Burritt. He was training to become a blacksmith when his attention shifted to astronomy and mathematics. He would have been around 18 years old at this time, which points to the Great Comet of 1811 as a possible source of inspiration.

The Burritt family struggled financially but in 1816, thanks to assistance from church friends, Elijah entered Williams College in Massachusetts. To help pay for college, he took a temporary teaching job at Sanderson Academy in nearby Ashfield, the birthplace of renowned telescope maker Alvan Clark. Burritt returned to Williams College in 1818 and also wrote his first book, *Logarithmick Arithmetick*, which emphasized astronomical calculations.

The following year, Burritt moved to Milledgeville, Georgia, where he would utilize his maths and science skills as a teacher. He soon married Ann Watson and started a family. A gifted writer, he ran a weekly newspaper called the *Statesman and Patriot* and wrote "Astronomia," a 28-page introduction to the night sky.

Slaves comprised about half the population of Milledgeville, then the state capital. City

Miniature hand-painted portrait of Elijah Hinsdale Burritt. (Elihu Burritt Library, Central Connecticut State University).

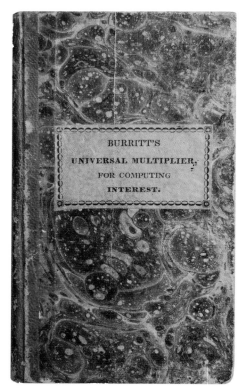
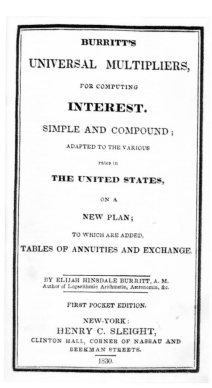

The 72-page "pocket edition" of *Burritt's Universal Multipliers* is one of several 1830 versions of this popular work.

politics was dominated by two warring factions and newspaper editors were forced to choose sides. A decade after moving to Georgia, Burritt suddenly found himself in mortal danger.

In 1829, an African-American abolitionist from Boston named David Walker wrote a pamphlet titled *An Appeal to the Coloured Citizens of the World*. Walker's pamphlet eloquently encouraged the Black community to fight slavery and its appearance triggered a wave of fear and anger in the antebellum South. Georgia's legislators reacted by passing a law making the distribution of abolitionist literature to the slave population a capital offence.

Elijah Burritt was eager to read and report on the pamphlet and wrote to Walker asking for a couple of copies. Walker sent Burritt twenty pamphlets, together with a letter dangerously implying that the two were friends and possibly even conspirators.

Burritt gave copies of "An Appeal" to a handful of curious associates, including his co-editor, John Polhill, and stored the remainder in his office. Polhill became suspicious when he intercepted a second letter from David Walker while Burritt was out of town. He met with a small group of community leaders and they passed what they believed was incriminating evidence on to the governor.

Burritt was twice arrested and released on legal technicalities. Fearing that his life was in jeopardy, he fled back to New Britain, Connecticut, just ahead of a grand-jury indictment. Despite being a Northerner, Burritt was not an abolitionist and had broken no laws in Georgia. Historian Glenn M. McNair has shown that Burritt's only transgressions were his support of tariffs and his failure to provide vigorous partisan support within the pages of his newspaper, both of which invited the wrath of his political allies. His interest in Walker's pamphlet provided the perfect opportunity for a group of powerful Georgians to orchestrate a politically-motivated takeover of his newspaper by exploiting abolitionist paranoia.

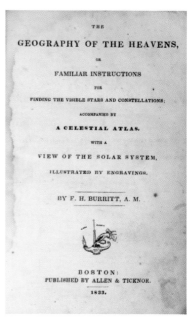

The title page of the 1833 second edition of *The Geography of the Heavens* describes the book as "familiar instructions" while later editions characterized the work as a "class book of astronomy."

Back in Connecticut, Burritt engaged in a letter-writing campaign as he and his wife attempted to clear his name in Georgia. He also purchased a building and transformed it into a schoolhouse, complete with a telescope, and opened a day and boarding school. During this hectic time, he wrote "Universal Multipliers," a 32-page booklet published in 1830 that provided useful information about interest calculations and annuities. He then sold the copyright to a wholesale book publisher in New York City for a sizable sum.

In a letter dated 12 November 1830, Burritt mentions an astronomical work "I may ere long publish," hinting that he probably devoted a couple years to creating his *magnum opus*, a user-friendly introduction to the night sky.

"I have long felt the want of a Class Book, which should be to the starry heavens, what Geography is to the Earth," wrote Elijah Burritt in his introduction to *The Geography of the Heavens*. This guidebook appeared in February 1833 and

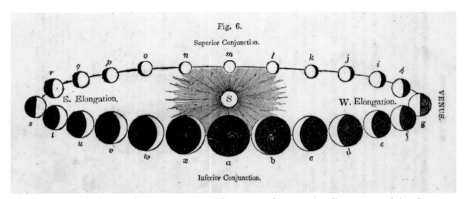

The 1833 second edition of *The Geography of the Heavens* features this illustration of the changing phases of Venus. "Thus is the system of Ptolemy weakened," remarked Burritt, "while that of Copernicus receives confirmation."

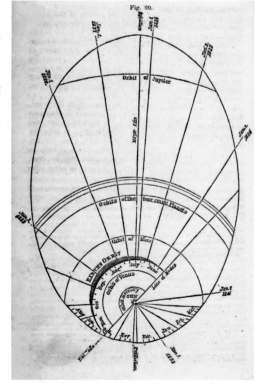

featured descriptions and histories of the constellations, together with information about the brighter stars and other objects. Burritt had perfected his interpretive skills during nighttime stargazing programs for students and this experience is reflected in the book's lucid descriptions.

The book was accompanied by *Atlas Designed to Illustrate the Geography of the Heavens*, which divided the night sky into six large maps showing stars down to sixth magnitude. It featured mythological figures, coloured by

The orbit of Biela's Comet, the second comet known to have a very short period, is depicted in *The Geography of the Heavens*. The comet had been recovered by John Herschel in 1832, around the time that Burritt was putting the finishing touches to his book.

Burritt's *Atlas Designed to Illustrate the Geography of the Heavens* received a complete makeover prior to the publication of the 1835 edition. The "Northern Circumpolar Map," pictured here, features improved mythological figures.

hand, which were based on Francis Wollaston's 1811 atlas. Burritt's guidebook and atlas were designed to "supersede the necessity of celestial globes in schools," since globes were expensive and confusingly depicted the constellations backwards.

The first edition of *Geography* was received with enthusiasm and sold out quickly. A second edition followed in September 1833. Responding to feedback from teachers, Burritt expanded his description of the solar system. The page-count jumped from 264 to 342 as the book evolved from an atlas companion into an astronomy textbook. A solar system illustration spanning two pages was added to the 1835 edition of the atlas, which also featured newly-engraved star maps

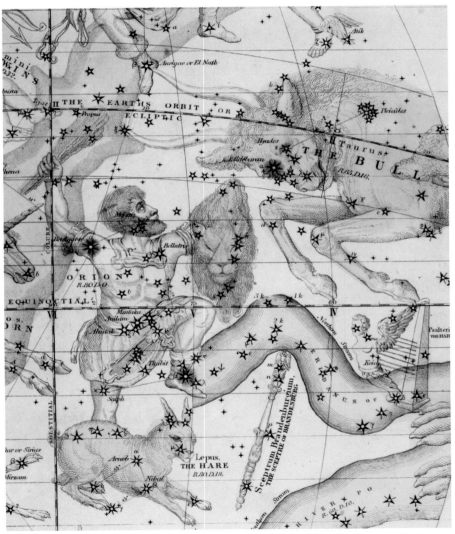

The beauty of Elijah Burritt's star atlas is enhanced by the laborious hand application of watercolour paint, as illustrated in this detail from the 1835 edition.

with bolder and more artistic mythological figures that were based on the work of Scottish author Alexander Jamieson.

In 1836, the same year that a third edition of Burritt's *The Geography of the Heavens* was published, Texas seceded from Mexico and became an independent republic. Brothers John and Augustus Allen, who were establishing a new town named Houston, convinced 43-year-old Elijah Burritt to assemble a group of craftsmen and bring them to Texas to form the Texas Steam Mill Company.

"The commencement of this Texas Expedition has been full of toil and sacrifice," wrote Burritt to his wife as he departed from Connecticut, never dreaming that his ultimate sacrifice would come in just a few months. After a month-long voyage along interior waterways, Burritt's group arrived at Galveston on 29 September 1837, just ahead of a slow-moving hurricane that devastated the coastline. They managed to reach Houston, but overcrowded conditions forced them to live in tents. An outbreak of yellow fever soon claimed the lives of six group members including their energetic leader, Elijah Burritt, who died on 3 January 1838.

Burritt's life was cut short, but his passion for astronomy lived on through his famous guidebook and star atlas, which introduced many thousands of people to the magnificent beauty of the night sky. *The Geography of the Heavens* was so popular that new printings appeared almost yearly until well after the Civil War, with 300,000 copies in print by 1866. Even today, these quaint publications are coveted by collectors and remain an enduring legacy of Connecticut's geographer of the heavens.

The author wishes to thank Renata Vickrey, archivist at Central Connecticut State University, for providing copies of Elijah Burritt's letters.

The Future of Spaceflight

Andrew P. B. Lound

When I was five years old the first book that caught my attention on the classroom bookshelf was *The Rocket*, a Ladybird book published in 1967 and filled with images of rockets and how they work. The last image in the book had a profound effect upon me; it was of a boy and girl viewing the Moon from orbit looking out of a curtained window. I always thought that when I became as old as my father, I would be making that voyage. Of course it never happened, although it may be not too far away. Spaceflight has changed beyond all recognition since the late 1960s – the 'future' of spaceflight can no longer be counted in decades but in single years.

Children look out of a window onto the lunar surface in this illustration from the Ladybird book *The Rocket*. (Wills & Hepworth Ltd / Ladybird Books)

Generally speaking, spaceflight has been motivated by political aims and funded by a nation's treasure. The original Space Race between the Soviet Union and the United States led to staggering technological strides culminating with the Apollo Moon landings. Once the aims were achieved a fairly uncoordinated and ad hoc approach was taken to explore the universe and to develop technologies which were often repurposed military technology. The future of spaceflight is going to be different, for the nation states have a new competitor – private industry.

A generation of people who grew up with the excitement of space exploration and who perhaps, like me, wanted the image in the Ladybird book to come true, were disappointed. Some of these people became billionaires, such as Elon Musk, Jeff Bezos and Sir Richard Branson. They have used their wealth to form their own space programmes which are available for hire to governments, universities and in fact to anyone who can afford to pay. The benefits of the private sector becoming involved are that the aim to reduce the cost per kilogramme to orbit (and even sub-orbit) literally opens up the new frontier.

In the coming decade commercial spaceflights to low Earth orbit will transform space activities as, not only will the major nations be participating, but smaller nations, wealthy individuals and corporations will be able to afford to launch satellites, research projects and even people. Richard Branson with his Virgin Galactic company was the first to attempt to exploit an interest in sub-orbital flights using a reusable aircraft. Jeff Bezos with his Blue Origin system can either make sub-orbital flights – like a traditional rocket – or achieve low Earth orbit. Elon Musk's SpaceX developed the reusable Falcon rocket system to reach low Earth orbit, and the larger Falcon Heavy rocket to attain high Earth orbit, or beyond to the planets. The key to these companies is reusability – that is the game changer. This reduces costs which, along with the standardization of components and systems, makes the whole launch system far more practical than any of the previous

SpaceX Starship – designed to carry up to 100 people to the Moon and beyond. (SpaceX)

NASA human spaceflight projects. The push by private companies has led to the traditional aerospace companies such as Boeing developing their own spaceflight systems that are available for hire. The reduction in cost has left the expensive non-commercial research to governments, who will simply hire the launch system that best suits them. NASA's Volatiles Investigating Polar Exploration Rover (VIPER) will be sent to the Moon in 2023, launched on a commercial rocket and deployed by a commercial lander. This will be the future model.

The privatisation of spaceflight in the West will lead to commercial space stations that will house research astronauts as well as tourists. SpaceX also has its Starship spacecraft. A truly remarkable development, it utilises a rocket far more powerful than the Saturn V that launched men to the Moon. Starship has been designed to carry humans to the Moon, Mars or even beyond. All it needs is refuelling, and SpaceX has a refuelling system in hand. In essence, the SpaceX system is a spacecraft that can be used for whatever the customer requires – a true interplanetary vehicle once only the realm of science fiction writers.

Artist's impression of activities at the Artemis Moonbase – America's outpost on the Moon. (NASA/ESA/P. Carril)

In the East, China, Russia and India are maintaining a more traditional approach to funding space activities, but the goals of co-exploitation are the same as in the West. With the cost to launch falling (SpaceX's Starship is estimated to be $2 million per launch) the race to exploit resources in space is now a commercial possibility.

With resources on the Earth becoming ever harder to obtain – deep sea drilling is costly and mines are becoming more inaccessible – it will be cheaper to mine the Moon and the asteroids for the materials needed for electronics, power and the chemical industry. So, within the next thirty years or so, mineral and raw material exploitation of space will become more prevalent.

Space tourism will naturally increase, and actually goes hand in hand with the exploitation of resources. Lunar bases may be built by nation states, or at least come under the jurisdiction of nation states if they have been built by private concerns. Setting up mining operations is expensive, so to pay for the construction some modules will be made available for wealthy tourists or companies. This would create a new tourist industry in much the same way as when commercial flying started, first for the wealthy and then later for the general public.

I can foresee the Moon looking very much like what was envisioned in the movie *Moon Zero Two* (1969). Although this was tongue in cheek, the basic elements of the movie are coming true.

This is, of course, the expensive exotica, but back here on Earth spaceflight is a necessity for the future of travel. In Britain, the Oxfordshire-based company Reaction Engines Limited is developing Skylon, a fabulous spaceplane with air breathing engines that can reach orbit from a long runway and return. The technology of Skylon is vital for future travel. As the planet struggles with global warming, cleaner fuels will be required, and are likely to involve hydrogen either as a gas or

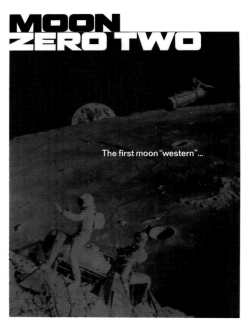

Poster for the movie *Moon Zero Two* which depicts the Moon as a wild frontier. (From the author's collection)

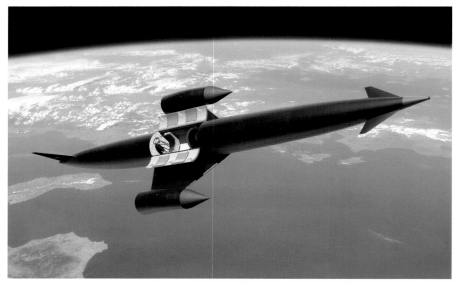

Skylon – a British designed spaceplane. (Reaction Engines Ltd)

liquid. Skylon uses liquid hydrogen which means the only exhaust waste is water vapour. But it is not simply the fuel. In the skies at any given day there are nearly 10,000 planes carrying around one and a quarter million passengers. If we could reduce the number of aircraft flying and still transport passengers, it would be a huge environmental benefit. Skylon's technology offers the opportunity to develop sub-orbit passenger spacecraft. Such flights would reduce the journey times across the world and would transform the global economy. One could literally commute from London to Sydney in Australia on a daily basis. Spaceports are already planned for the UK with simple small payload rocket launches being the main activity in the short term, but long term Skylon or its derivatives will take off from the UK.

The future of spaceflight as we can envisage it today is one of mixed commercial and state sponsored missions. With it will come a whole new era of human activity. 'Space: the final frontier' will literally become a wild frontier. Private companies may vie with each other for rights to raw materials on the Moon and nation states may become embroiled in these disputes. International space law is currently under the regulations laid out in the United Nations Outer Space Treaty 1967. In this document no nation or organization may lay claim to any territory of a celestial body. How this treaty can be enforced becomes an issue. China has recently revealed that its rover on the far side of the Moon has detected Helium 3,

Sub-orbital passenger plane – a concept by Reaction Engines and the UK Space Agency. (Reaction Engines Ltd)

an isotope of Helium not found on the Earth, but material that would be vital for future nuclear fusion reactors. Helium 3 could form the basis of a new Klondike on the Moon. As in the old western movies, prospectors could be setting up small bases in which to locate and extract raw materials. In those Wild West days who was the law? The 1967 treaty is in desperate need for updating, with understandings of legal jurisdiction.

The United States has created a Space Force – mainly to safeguard American interests in orbit – although sooner or later this will extend beyond the orbit to protect US nationals in space. Naturally other nations will reciprocate. The future of spaceflight offers wondrous possibilities for discovery – the creation of a space faring species reaching out into the cosmos. However, with that species will come all of its frailties.

The near future will kick start the new 'outward urge', as Arthur C. Clarke referred to it. The United States will land people on the Moon as part of the Artemis program, certainly before 2030 as the 2024 date seems to have slipped. We may see circumnavigations of the Moon by a private firm before then – more likely SpaceX. Either way, unlike Apollo, the momentum will not stop. The calling of the other planets will be just too strong, especially as the technology will be available to hire by all nations. Mars before 2040, and then if the technologies work well, humans venturing further afield. And while commercial and national explorations venture forth, back on Earth passengers will make the daily commute to Sydney or

Orbital Hotels will take the form of a modular design. (Bigelow Aerospace)

New York from London. Tourists will enjoy the views of their home planet from orbiting hotels, and on the Moon they will play games in the sixth-gravity, and take photos with their super smart phones of Tranquillity Base – the memorial to Apollo 11.

Male Family Mentors for Women in Astronomy
Caroline and William Herschel

Mary McIntyre

When we talk about male mentors for female astronomers, the most well-known pairing has to be William and Caroline Herschel. They had an exceptionally impressive career together, dedicating much of their lives to astronomy and Caroline went on to become the first women ever to be paid a wage for her work in science.

Caroline Lucretia Herschel was born in Hanover in 1750, the eighth child to parents Isaac and Anna. Her father was a professional oboe player and music teacher. He educated his children in music but also gave them a basic grounding in many other subjects. Caroline contracted typhus aged ten and also suffered from smallpox in her childhood. These illnesses, coupled with childhood

Lithograph of Caroline Herschel in 1847, created when she was about 97 years old. (Wikimedia Commons/Unknown Author)

neglect at the hands of her mother, are thought to have stunted her growth, so she was only four feet and three inches tall. She also suffered vision loss in her left eye and this, combined with her petite stature, meant that her family believed she would never marry.

Caroline did not have a happy childhood. Her older sister Sophia married and left home when Caroline was five years old so being the only daughter still living at home, her mother condemned her to a life of domestic servitude. With the exception of her father and brother William, most of her family seemed oblivious to her unless she was being whipped or ridiculed. In her memoirs she stated that *"there was nobody who cared anything about me"*. She recalls a time when she was six years old and was sent out to look for and greet her father and brother William as they were due to return home. She searched endlessly in the extreme cold but couldn't find them. When she eventually returned home, bitterly cold and tired, she found the whole family (including Isaac and William) sitting around the table enjoying a meal. William leapt up to greet his *"dear Lina"* (William's pet name for Caroline) but nobody else had even noticed she was absent.

While still as young as seven years old she was sometimes expected to work as a footman or waiter. Being so young, she did not have the dexterity to perform these tasks well and she received many whippings for being *"too awkward"*. She was also tasked with sewing tents and linen for the army after school. There were no thimbles small enough for her delicate hands so whilst doing this she literally worked her fingers to the bone.

A dress worn by Caroline Herschel displayed at the Herschel Museum in Bath. It was worn when she was about 50 years old. (Wikimedia Commons/Mike Peel **www.mikepeel.net**/Herschel Museum of Astronomy/CC BY-SA 4.0)

Caroline's mother strongly disapproved of Isaac teaching Caroline and did everything she could to prevent Caroline from gaining education in any subject that would give her the skills to work as governess or indeed any other job that would allow her to become an independent woman. Anna wanted to keep her free domestic servant at home and Caroline was forced to work such long hours that she scarcely would have had time for lessons even if she had been permitted to take them.

Caroline's initial interest in the night sky came not from William, but from her father. On 1 April 1764, when Caroline was 14 years old, Isaac gathered his family around a tub of water so they could all safely observe a total solar eclipse. He also took Caroline out into the street to look at the night sky, and she recalls observing *"some of the most beautiful constellations"* and also *"a comet"*. Her father died in 1767 leaving Caroline in despair at her lack of any prospects for her future.

Caroline absolutely adored William and he cared so much for her that he wanted to get his *dear Lina* away from her miserable life. Her brothers William, Alexander and Jacob were working as professional musicians in Bath, and in 1771 William wrote to suggest she join them so he could teach her how to sing and allow her to have a music career. It's hard to imagine just how thrilled Caroline was to be given this opportunity, even though Jacob ridiculed her for it. Once she arrived in Bath she was frustrated again because William was just too busy to teach her. She became a housekeeper there and spent many hours alone teaching herself how to sing. She recounts how anxious she

felt having to go out to the market to buy food, with very little grasp of the English language and also concerned about the price of everything.

She did go on to have a successful music career, but as William's time became increasingly taken up by astronomy, he trained her to become an assistant

Institute of Physics plaque located at the rear of the Herschel Museum, Bath. This house was once home to William and Caroline Herschel. (Wikimedia Commons / Mike Peel **www.mikepeel.net** / Herschel Museum of Astronomy / CC BY-SA 4.0)

Caroline Herschel's telescope used for "sweeping" the sky for comets. It was made for her by William Herschel. (Wikimedia Commons / Geni / CC-BY-SA GFDL)

astronomer. Although Caroline desperately wanted to be an independent woman, her first priority was always William. She assisted his methodical observations of the sky with his home-made telescopes, planning their observing sessions and taking notes dictated by William from his observing position. The following day she would write up all of the previous night's observations and work through calculations. She also helped to build new telescopes by grinding and polish mirrors, replacing tarnished mirrors and mounting telescopes. They even had a furnace so they could cast their own mirrors. This was quite dangerous work and both of them sustained many injuries, including one occasion when part of Caroline's thigh muscle was torn away when she fell onto a large hook.

In 1782 they moved from Bath to Datchet (near Windsor) so that William would be on hand to entertain royal guests. Caroline was very unhappy living there, describing the house as being a *"ruin of a place"* and William was so busy that Caroline felt alone and isolated. Eventually however, in 1886 they moved to more suitable accommodation in Slough.

With Caroline's help, William went on to discover over 2,400 objects including the planet Uranus. In addition to helping William, Caroline was an exceptional astronomer in her own right. William built her a telescope of her own and on

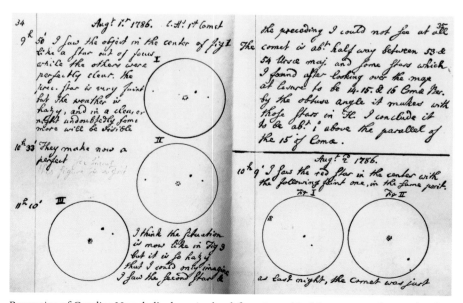

Recreation of Caroline Herschel's observing book from 1st and 2nd August 1786, showing her first comet discovery. (Mary McIntyre based on an original held at the Royal Astronomical Society)

The Andromeda Galaxy (M31) with satellite galaxies M32 and M110. M110 was excluded from Charles Messier's list even though he had first observed it in 1773. Caroline Herschel independently discovered it in 1783. (Mary McIntyre)

the nights that she wasn't required to help him, she would use it to "sweep" the sky looking for comets. She often had her telescope close to where William was working so she could be on hand at a moment's notice if he needed her help. She worked tirelessly at astronomy whilst still being a housekeeper and *"keeping the purse"*.

In 1783 Caroline made her first discovery – Messier 110, the companion galaxy to the Andromeda Galaxy – and in 1786 she discovered her first comet. Caroline is often credited for being the first female astronomer to discover a comet, but in fact, 81 years earlier Maria Kirch had discovered a comet (which was officially attributed to her husband Gottfried). Caroline went on to discover eight comets in total and worked on re-writing and making improvements to Flamsteed's Star Catalogue. In 1787 she became the first women ever to receive a salary for working in science when King George III gave William £2,000 towards the building costs of his 40-foot reflector, and added a provision that Caroline was paid a salary for her valuable work assisting her brother.

She was still working as William's assistant when on the 8 May 1788 William got married. Caroline had to give up her position of housekeeper and move to new lodgings which must have been an extremely difficult time for her. In her memoirs Caroline talks about how she found it very frustrating not having instant access to the library and observing notebooks when she was working at home. Her feelings towards the marriage itself at that time are not known because she destroyed many of her journals in later life, however from the letters she exchanged with her sister-in-law Mary, it is clear that they were extremely fond of each other and Caroline absolutely adored her nephew John.

In the final year of his life William was suffering with ill health. Caroline believed this was a result of the many years of constant sleep deprivation and being overworked, and in her memoirs she states that they would scarcely have had any sleep had it not been for cloudy nights or nights affected by the Full Moon. When he was unwell, Caroline would spend most afternoons visiting

Statue of William and Caroline Herschel in the garden of the Herschel Museum, Bath. (Wikimedia Commons/Mike Peel **www.mikepeel.net**/Herschel Museum of Astronomy/CC BY-SA 4.0)

him and once home she would forego her own rest time to work tirelessly writing up findings, memoranda, working on the star catalogue and copying William's many papers that had been published.

William died on 25 August 1822 and Caroline was, understandably, utterly grief-stricken. She very quickly made arrangements to move back to Hanover, believing that she wouldn't live longer than a year; a decision she later regretted. Once back in Hanover Caroline was unable to carry out any visual observing – another thing that she found incredibly frustrating – but she continued working to verify and write up all of their previous observations as well as working on the star catalogue. She did this despite her failing eyesight.

It was during her later life that she began to receive many accolades, including being awarded the Royal Astronomical Society Gold Medal in 1828, an honour that was not given to a woman again until 1996, when it was awarded to Vera Rubin.

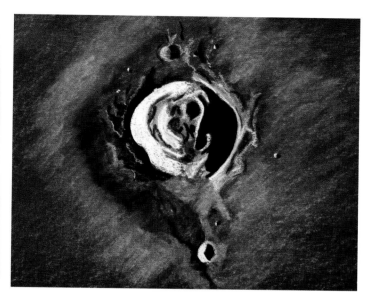

A pastel sketch of the lunar crater C. Herschel. This crater is located on the western part of Mare Imbrium and cuts across the wrinkle ridge Dorsum Heim. (Mary McIntyre)

In 1835 she was elected as an Honorary Member of the RAS and in 1846 she was awarded a gold medal for science by the King of Prussia.

Caroline did not feel deserving of any of the awards she received, partly because she wasn't actively observing any longer, but also because she felt any praise for her work was diminishing William's achievements. In her memoirs she says *"I did nothing for my brother than what a well-trained puppy-dog would have done"*. This is simply not true! Reading the correspondence between Caroline and other astronomers of the time, including her nephew John Herschel, it is clear she was held in extremely high esteem by everybody who knew her or knew of her.

Caroline died peacefully in 1848 at the age of 97. She was buried next to her parents and with a lock of William's hair. The lunar crater C. Herschel, open clusters NCG 2360 (Caroline's Cluster) and NGC 7789 (Caroline's Rose), and asteroid 128 Lucretia are all named in her honour. In 2020, the Argentinean Earth observation satellite ŇuSat 10 "Caroline" – named after her – was launched into space.

I wish that Caroline could see herself through our eyes; she was a truly remarkable and courageous lady whose grit and determination to succeed despite the challenges she faced in life, truly changed the landscape for all women who followed in her footsteps.

The third and final part of our Male Mentors for Women in Astronomy series will be published in the *Yearbook of Astronomy 2024*.

The Closing of Historic Observatories

Harold A. McAlister

Closing Domes

George Ellery Hale – founder of the Yerkes, Mount Wilson, and Mount Palomar observatories – was driven to create ever-larger telescopes located at the best available sites, an ambition that has only accelerated with time. Hale's Mount Wilson 100-inch telescope was the world's largest from 1917 until 1949 when his final creation – the Palomar 200 inch – became operational. The latter then held the record for the next 26 years before it was trumped by the Soviet 236-inch telescope. With the development of cost-saving technologies involving such things as active mirror supports, adaptive optics, and spun-cast and segmented mirrors of short and compact focal length, ever larger

The dome of the Mount Wilson 100-inch reflector photographed by the author in October 2013. (Hal McAlister)

telescopes became more affordable. The twenty-first century has seen a plethora of new telescopes of effective apertures in the 240- to 470-inch range. Presently in the works are telescopes as large as 1,550 inches, a stupendous aperture yielding a light-gathering power 240 times that of Hale's 100 inch. To turn that gain around, the Mount Wilson telescope must be compared to a 6.5-inch instrument. Couple such gains with the factor of 30 or so increase in quantum efficiency of modern detectors in comparison with the best photographic emulsions, we see that when the European Extremely Large Telescope (ELT) is built it will record the same number of photos in 6 seconds as did the 100-inch telescope in an exposure lasting a long winter's night. And that's not all. The ELT's adaptive optics will produce images 16 times sharper than those from the Hubble Space Telescope. Hale would have loved this!

And yet, it is an exquisitely moving experience to stand on the observing floor of the Mount Wilson 100-inch telescope dome at dusk watching and listening to the huge shutters slowly opening to give the giant telescope, looming before you, access to the darkening sky. One has the sense of standing in a cathedral of science – a place that should endure forever. After all, an astronomical telescope is only as obsolete as the equipment one fixes to the instrument's focal plane. So, there is no need to ever stop opening that grand dome on clear nights. In fact, the 200-inch telescope is still operating under this premise.

But, of course, these "lighthouses of the sky," as the sixth US president John Quincy Adams dubbed them, require substantial funding to

Inside the 100-inch dome, the shutters having been opened for a night's work. Photographed by the author in April 2013. (Hal McAlister)

operate and maintain, and that support can vanish for any number of reasons. In 1904, George Ellery Hale persuaded the newly-founded Carnegie Institution of Washington (CIW) to establish an observatory in the San Gabriel Mountains of Southern California that would revolutionize telescope design and launch modern observational astrophysics. In doing so, Hale walked away from Yerkes Observatory that he had founded only seven years earlier, taking many of its staff to his far better site at the western edge of the continent. Hale would eventually look 100 miles south eastward from Mount Wilson toward Palomar Mountain as a darker site for which he and his staff would design the 200-inch telescope.

With the population of Los Angeles County increasing annually by 100,000 inhabitants during the twentieth century, the night sky illumination from city lights grew relentlessly. To ensure continuing access to the faintest possible objects, the Carnegie Institution had established its Las Campanas Observatory beneath the very dark Andean skies of Chile and was operating a modern 98-inch telescope there by 1977. Much more was envisioned for Las Campanas. In June 1984, CIW announced that it would cease funding Mount Wilson operations, effective 1 July 1985, in order to redirect observatory funding primarily to its Chilean operation. The dome of the observatory's crown jewel – the 100-inch telescope – would cease its grand nightly openings.

The owners were hopeful that some new entity would arise and provide new scientific and fiscal management of Mount Wilson, but it was clear CIW would be unswayed by emotionally-based or historic-preservation arguments. Any successor organization would only be approved if it proposed continuing worthwhile science from Mount Wilson. In any event, Carnegie would step away from Mount Wilson following their 81st and final year of operations.

The 200-inch mirror blank arriving in Pasadena on 10 April 1936 as photographed by the author's mentor Karel Hujer. (From the author's collection)

Mount Wilson's Rescue

Through the efforts of JPL astronomer Arthur H. Vaughan, attorney C. Robert Ferguson, and Smithsonian astronomer Robert W. Noyes, the Mount Wilson Institute (MWI) was formed consistent with CIW's goals. On 5 January 1989, Carnegie signed over the Observatory's operations to MWI with Vaughan acting as Mount Wilson's director. In 1992, Vaughan was succeeded by Robert Jastrow, famed as a populariser of astronomy. Jastrow, who, like Vaughan, served as CEO and Director in a pro bono capacity, energetically set about raising funds to modernize the 100-inch telescope's drive and control system and to equip it with a state-of-the-art adaptive-optics system. He then embarked on an if-you-build-it, they-will-come business model of selling observing time to astronomers that would regrettably prove to be inconsistent with science funding realities, and the venture was not successful.

Having selected Mount Wilson as the site for the Center for High Angular Resolution Astronomy (CHARA) Array project in 1996, I soon became a member of the MWI board. When Bob Jastrow announced his retirement in late 2001, I agreed to undertake the MWI directorship on top of my duties as director of Georgia State University's long-baseline interferometric telescope array. I felt that Mount Wilson's future lay elsewhere than in the scientific utilization of its original facilities, which I regard to be of world-heritage class. Mount Wilson could become a kind of Colonial Williamsburg of astronomy. Its situation in Southern California imparts proximity to millions of potential visitors. With the collaboration of Pasadena architect Stefanos Polyzoides, ambitious plans for a large visitor centre were developed. However, the MWI board of trustees felt my goals were over-reaching, and funding for the visitor centre was never pursued.

In 2014, I stepped down as director – the last professional astronomer in that role. Today, MWI derives its income from site fees charged to mountain tenants such as CHARA, sales of tour tickets, public observing on the 60- and 100-inch telescopes, summer concerts in the 100-inch dome, and the occasional serendipitous bequest. This provides funding for a small maintenance crew, but there remain no salaried management staff. In spite of not having an endowment, MWI has managed to operate the mountain for more than three decades, taking Mount Wilson Observatory well into its second century.

The Yerkes Transition

Hale's first great observatory would also encounter financial realities. The University of Chicago's Yerkes Observatory is located in Williams Bay, Wisconsin, some 70 miles north west of its parent campus. According to Kyle Cudworth, director of Yerkes Observatory from 2001 to 2012, a University of Chicago (UC) president had declared as early as 1970 a desire to close Yerkes and redirect the facility's funding to higher priorities. In the mid-1980s, an attempt was made to sell the nearly 80 acres surrounding the Observatory on upscale Geneva Lake to a developer, but it became clear that the required rezoning would be denied. The status quo would be maintained for another few decades. During that time the interests of the UC astronomy faculty had largely drifted away from stellar to extragalactic astronomy and cosmology, and the Yerkes telescopes became mostly irrelevant to them. Graduate students stopped spending extensive periods of time in Williams Bay, and grant funding to carry out science at the 40-inch refractor – still the world's largest lens-based telescope – became increasingly limited. Responding to these changes, Cudworth and other Yerkes devotees switched their focus to education and public outreach and obtained external funding to sponsor these activities. However, in

The architectural and materials aspects of the dome of the Yerkes 40-inch refractor, imaged by the author in September 2007, contrast strongly with the utilitarian design and construction of the 100-inch telescope dome. (Hal McAlister)

spite of these new resources, the core operating expenses still fell to the University of Chicago.

In the spring of 2018, the inevitable became reality, and UC announced it would cease Yerkes operations effective 1 October 2018. The University invited proposals from interested parties to assume full responsibility for Yerkes. The Yerkes Future Foundation (YFF) was quickly formed by a group of influential residents of the Geneva Lake area who submitted a plan to fund and manage the Observatory. Following lengthy negotiations, the University entered into an agreement with the YFF who accepted a donation of the facility and its surrounding 49 acres in May of 2020. The Foundation envisions a continuing role in research and education. After raising several million dollars in the first few months of its capital campaign, the new owners have appointed a new executive director experienced in managing cultural institutions while investing substantial resources in badly needed maintenance for the architecturally magnificent 125 year-old building. It seems likely that Yerkes' future is in good hands and the Observatory will once again be operating in some limited extent prior to the appearance of this article.

A Close Call

Lick Observatory is another famous American astronomical site threatened with closure. The observatory's development began 20 years prior to Yerkes' opening, at which point Lick's 36-inch telescope surrendered its "largest refractor" status. Soon after its completion, Lick was turned over to the University of California where it has served astronomers from multiple campuses for well over a century. Located 20 miles east of San Jose, California, atop Mount Hamilton at an elevation of 4,265 feet, Lick's climate is considerably better suited to astronomy than Yerkes'. By 1959, a 120-inch reflector – surpassed in size for years only by the Palomar 200 inch – became operational on Mount Hamilton, markedly enhancing the facility's research capability. However, Silicon Valley light pollution inevitably presented problems to Lick. That encroachment – along with financial considerations – led to reductions in Mount Hamilton staff beginning around 2008. In 2013, it was announced that a major funding source would dry up over the next five years, presaging the observatory's closure. That announcement was rescinded a year later when UC asserted that operations would continue, and Google subsequently donated $1 million to support Lick. The Observatory appears to be operating stably and unthreatened in the near term.

The Lick Observatory's 36-inch refractor is seen in its stow position in this July 2013 image by the author. (Hal McAlister)

Reinventing an Observatory

Far to the north east of Lick is the David Dunlap Observatory in Richmond Hill, Ontario, 15 miles north of downtown Toronto. Opening in 1935, the University of Toronto's DDO boasted a 74-inch telescope, second in size only to the Mount Wilson 100-inch. Toronto astronomers and their students effectively used the instrument for decades with an emphasis on observational stellar astrophysics. In 1971, DDO spectroscopy showed that the companion of Cygnus X-1 was most likely a black hole. By the 1960s, the northward spread of metropolitan Toronto had brightened the night sky in Richmond Hill. Rather than relocate the 74-inch to a new dark site, it was decided to provide for a modern 24-inch telescope to be hosted by CIW at Las Campanas. By 1998, Canadian astronomy looked toward international partnerships to enable access to very large telescopes and cutting-edge

University of Toronto astronomer Thomas Bolton, who passed away in 2021, stands inside the DDO 74-inch telescope in this ca.1973 photograph around the time that he was playing a key role in demonstrating that Cygnus X-1 is a black hole binary system. (Courtesy Ian Shelton)

instrumentation. The 24-inch was relocated to Argentina with Toronto keeping only a minor share of its time. Their divestiture from wholly-owned facilities was a strategic move, and in 2007, UT announced its plans to sell DDO and its surrounding 175 acres with a portion of the proceeds creating an endowment of an astronomical institute. The sale for $70 million occurred the following year. The development details have proven to be a complex undertaking involving a number of interested parties, and emotions ran high. For the present, DDO is managed by the city of Richmond Hill with programs jointly conducted with other organizations. However, the Dunlap Institute for Astronomy and Astrophysics, with a focus on astronomical instrumentation, was indeed established. Its affiliation with UT's Department of Astronomy and Astrophysics provides for a good ending to this particular story.

Tragic Endings and Close Calls

Other endings have not been so happy. The most recent example is the Arecibo Observatory whose 1,000-foot fixed main instrument was for half a century the largest single-dish radio telescope in the world. Among its many "firsts", Arecibo is famed for its radar imaging of Mercury and Venus, first discovery of an exoplanet system, and demonstration that the Crab Nebula pulsar is a neutron star. Following breaks in the receiver platform cables in late-2020, the U.S. National Science Foundation (NSF) decided to demolish the telescope, a fate cut short by the platform's spontaneous collapse on 1 December 2020. While there is extensive justification for a rebuild of a modern 1,000-foot telescope, there is at present no commitment to a successor to this remarkable instrument.

This pyrocumulus cloud – produced by the intense heat from a burn area – erupted near Mount Wilson during the Station Fire on 4 September 2009. (Susan McAlister)

Another dramatic end of a major historic observatory occurred on 18 January 2003, when a firestorm swept through an area near Canberra, Australia destroying five telescopes and other facilities of the Mount Stromlo Observatory of the Australian National University (ANU). The site is now home to the ANU's Research School of Astronomy and Astrophysics, a major centre for instrumentation development and a partner in the Giant Magellan Telescope project – truly a phoenix rising from the ashes.

Other major observatories, typically built in remote and hard to defend locations, have been threatened by wildfires. For example, Mount Wilson has had several close encounters, the most recent of which were the August 2009 "Station Fire" and the September 2020 "Bobcat Fire." Just a month prior to the Bobcat Fire, Lick Observatory lost several residences to the "SCU Lightning Complex Fire." While the Station Fire was set by an arsonist, it is apparent that wildfire endangerment of observatories is being exacerbated worldwide by global warming.

Life, and Science, Goes On …

While great and historic observatories are timeless, their once huge telescopes – perhaps nearly blinded by light pollution – are dwarfed by their modern successors that exert great pressure to absorb their already diminished budgets and personnel. Some will be repurposed for outreach and educational usage following periods of consternation and anguish. Particularly historic observatories – such as the Royal Observatory at Greenwich – may become national museum sites. The Pulkova Observatory was rebuilt after its buildings were destroyed in WWII. Today, the institution is still in research use while also being a component of the St. Petersburg, Russia, UNESCO World Heritage Site.

Astronomical observatories may occasionally be abandoned or even demolished, but their grandeur transcends research facilities of the other natural sciences. Grassroots initiatives will continue to preserve historically important observatories for the enrichment of future generations. But light pollution, obsolescence, and finite budgets will also continue taking their toll. Such is in the nature of advancing science – just as it is in life, the old must make way for the new.

Miscellaneous

Some Interesting Variable Stars

Tracie Heywood

You may have considered taking up variable star observing but how should you choose which stars to observe? There are so many variable stars in the night sky and you don't want to waste your time attempting to follow the "boring" ones. Your choice of stars will, of course, depend on the equipment that you have available, but also needs to be influenced by how much time you can set aside for observing.

This article splits some of the more interesting variable stars into three groups. The group that is most suited to you will depend on how often you can observe each month and for how long you can observe on a clear night. The light curves included have been constructed from observations stored in the Photometry Database of the British Astronomical Association Variable Star Section. Comparison charts for most of these stars can be found on the BAA Variable Star Section website at **www.britastro.org/vss**

One-Nighters

These are stars that can go through most of their brightness variations in the course of a single (reasonably long) night and would suit *people who can only observe occasionally, but can then observe well into the night.*

STAR	TYPE	RA		DEC		MAX / MIN	PERIOD
		H	M	o	'		
AC Boötis	EW	14	56	+46	22	10.0 / 10.6	0.3524484 days (~8.5 hours)
VZ Cancri	Delta Sct	08	41	+09	49	7.18 / 7.91	0.1783638 days (~4.3 hours)
AI Draconis	EA	16	56	+52	42	7.0 / 8.0	1.1988146 days (~29 hours)
Beta (β) Persei (Algol)	EA	03	08	+40	57	2.1 / 3.4	2.867 days (~69 hours)

AC Boötis is a W UMa type eclipsing variable. These are 'contact' binary systems that show no period of constant brightness between eclipses. Predictions for upcoming eclipses can be found at **www.as.up.krakow.pl/minicalc/BOOAC.HTM**

VZ Cancri is a Delta Scuti type variable. Most variables of this type have brightness ranges of below 0.2 magnitudes and so cannot be accurately followed visually. VZ Cancri, however, varies over a range of around 0.7 magnitudes over a period of just 4.3 hours, making it one of the highest amplitude stars of this type.

AI Draconis is an Algol-type eclipsing variable star that is located near the 'head' of Draco. Predictions for upcoming eclipses can be found at **www.as.up.krakow. pl/minicalc/DRAAI.HTM**

Beta Persei (Algol) is an eclipsing variable that shows deep primary eclipses but only very shallow secondary eclipses. Although circumpolar from the UK, it is too low in the sky for observation from April to June.

Most Clear Nights

These are stars that vary a bit more slowly, but which can display significant changes over a week or two. They would suit *people who can observe for a short while on (nearly) every clear night.*

STAR	TYPE	RA		DEC		MAX / MIN	PERIOD
		H	**M**	**°**	**'**		
Z Camelopardalis	UGZ	08	25	+73	06	10.0 / 14.5	6.9562 hours (orbital)
Delta (δ) Cephei	Cepheid	22	29	+58	25	3.5 / 4.3	5.366266 days
V 742 Lyrae	RCB	18	38	+47	23	11.5 / 18.8	None
DY Persei	DYPer	02	35	+56	09	10.0 / 16.5	None

Z Camelopardalis is a dwarf nova system that sometimes shows 'standstills' that last for many months.

Delta (δ) Cephei is a classical Cepheid variable. As for other classical Cepheids, the brightness changes repeat exactly from one 10.15-day cycle to the next. Although visible with the naked eye, binoculars may be beneficial when making brightness estimates.

V742 Lyrae is a similar type of variable to R Coronae Borealis. It can show deep fades, unpredictable in advance, and these can last for many months.

DY Persei at first sight can appear to be a similar type of variable star to R Coronae Borealis. However, whereas RCB type variables are typically yellow supergiant stars, DYPer stars are cool carbon stars.

Several Times per Month

Slower variables whose brightness will change significantly over several months or a year. These variables would suit *people who can observe several times per month, but not necessarily on every clear night.*

STAR	TYPE	RA		DEC		MAX / MIN	PERIOD
		H	M	°	′		
W Andromedae	Mira	02	18	+44	18	6.7 / 14.6	397 days
Mu (μ) Cephei	SR	21	44	+58	47	3.4 / 5.1	835 days / 4,400 days
BU Geminorum	Irregular	06	12	+22	54	6.0 / 7.9	None
SS Herculis	Mira	16	33	+06	51	8.5 / 13.5	114 days
Y Lyncis	SR	07	28	+45	59	6.6 / 8.7	110 days / 1,400 days
BQ Orionis	SR	05	57	+22	50	7.1 / 9.0	243 days
R Ursae Majoris	Mira	10	45	+68	47	6.7 / 13.7	302 days

W Andromedae is a Mira type variable. As for other Mira type variables, some maxima are brighter than others. Recent maxima have ranged between magnitudes 6.8 and 9.7. The 2023 maximum is predicted to occur in late August.

Mu (μ) Cephei – Herschel's Garnet Star – is a semi-regular variable. "Semi-regular" indicates that the brightness variations repeat only roughly from one cycle to the next. The observed brightness range differs between cycles, sometimes covering the whole of the listed range, but usually being somewhat smaller.

BU Geminorum is an irregular variable. Recent studies have, however, suggested the presence of a period of around 460 days.

SS Herculis is a Mira type variable with an unusually short period of around 114 days. It is poorly placed for observation at the start of the year but better placed from the spring to mid-autumn. Maxima in 2023 are predicted to occur in late March, mid July and late October.

Y Lyncis and **BQ Orionis** are semi-regular variables, both of which often vary over most of their listed brightness range.

R Ursae Majoris is a Mira-type variable. It will start 2023 rising in brightness from minimum towards its predicted April maximum. It will then fade towards a mid-autumn minimum.

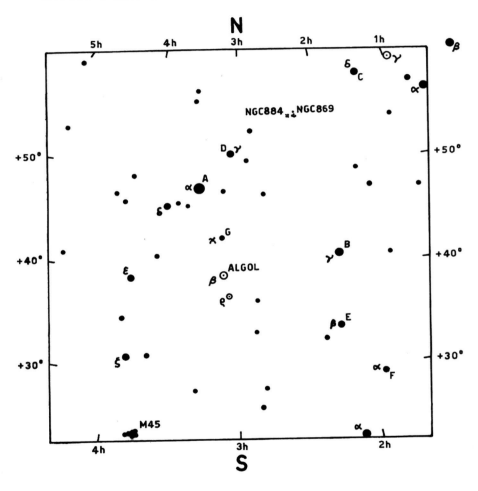

40° FIELD DIRECT

BETA PERSEI 03h 08m 10·1s +40° 57' 20" (2000)

CHART: A 1·8 E 3·0 BAA VSS
NORTONS STAR ATLAS B 2·1 F 3·4 EPOCH: 2000
  C 2·7 G 3·8 DRAWN: JT 19-06-11
SEQUENCE: D 2·9 APPROVED: RDP
HIPPARCOS VJ

The BAA VSS finder chart for Beta (β) Persei (Algol). (BAA Variable Star Section)

Minima of Algol in 2023

Beta (β) Persei (Algol): Magnitude 2.1 to 3.4 / Duration 9.6 hours

Month	Day	h		Month	Day	h		Month	Day	h		Month	Day	h	
Jan	2	3.7	⋆	Feb	2	16.7		Mar	3	8.9		Apr	1	1.0	
	5	0.5	⋆		5	13.5			6	5.7			3	21.8	⋆
	7	21.4	⋆		8	10.3			9	2.5			6	18.7	
	10	18.2	⋆		11	7.1			11	23.3	⋆		9	15.5	
	13	15.0			14	4.0			14	20.1	⋆		12	12.3	
	16	11.8			17	0.8	⋆		17	16.9			15	9.1	
	19	8.6			19	21.6	⋆		20	13.8			18	5.9	
	22	5.4			22	18.4	⋆		23	10.6			21	2.7	
	25	2.3	⋆		25	15.2			26	7.4			23	23.6	
	27	23.1	⋆		28	12.0			29	4.2			26	20.4	
	30	19.9	⋆										29	17.2	
May	2	14.0		Jun	3	3.0		Jul	1	19.1		Aug	2	8.1	
	5	10.8			5	23.8			4	16.0			5	4.9	
	8	7.6			8	20.6			7	12.8			8	1.7	⋆
	11	4.4			11	17.4			10	9.6			10	22.6	
	14	1.3			14	14.2			13	6.4			13	19.4	
	16	22.1			17	11.1			16	3.2			16	16.2	
	19	18.9			20	7.9			19	0.0			19	13.0	
	22	15.7			23	4.7			21	20.9			22	9.8	
	25	12.5			26	1.5			24	17.7			25	6.6	
	28	9.3			28	22.3			27	14.5			28	3.5	⋆
	31	6.2							30	11.3			31	0.3	⋆
Sep	2	21.1	⋆	Oct	1	13.3		Nov	2	2.2	⋆	Dec	3	15.2	
	5	17.9			4	10.1			4	23.0	⋆		6	12.0	
	8	14.7			7	6.9			7	19.9	⋆		9	8.8	
	11	11.5			10	3.7	⋆		10	16.7			12	5.7	
	14	8.4			13	0.5	⋆		13	13.5			15	2.5	⋆
	17	5.2			15	21.3	⋆		16	10.3			17	23.3	⋆
	20	2.0	⋆		18	18.2			19	7.1			20	20.1	⋆
	22	22.8	⋆		21	15.0			22	3.9	⋆		23	16.9	
	25	19.6	⋆		24	11.8			25	0.8	⋆		26	13.7	
	28	16.4			27	8.6			27	21.6	⋆		29	10.6	
					30	5.4	⋆		30	18.4	⋆				

Eclipses marked with an asterisk (⋆) are favourable from the British Isles, taking into account the altitude of the variable and the distance of the Sun below the horizon (based on longitude 0° and latitude 52°N)

All times given in the above table are expressed in UT / GMT.

μ mu CEPHEI   21h 43m 30·5s +58° 46′48″ (2000)

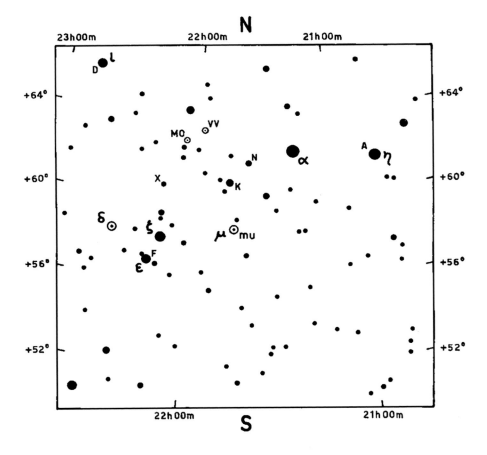

CHART: A 3·4   K 4·3   BAA VSS
SKY ATLAS 2000   D 3·5   N 4·8   EPOCH: 2000
SEQUENCE: F 4·2   X 5·4   DRAWN: JT 04-08-12
TYCHO 2 VJ   APPROVED: RDP

The BAA VSS finder chart for Mu (μ) Cephei, with six suitable comparison stars labelled. The same comparison stars can also be used for Delta (δ) Cephei, which is also labelled on the chart. (BAA Variable Star Section)

Light Curve for AC BOO

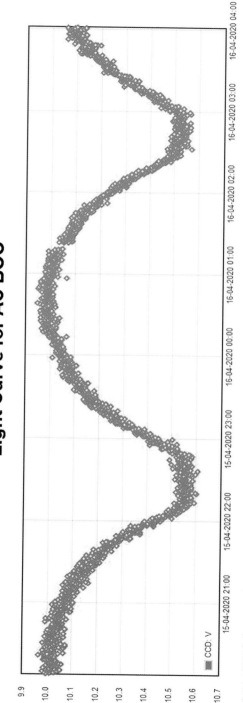

A BAA VSS light curve showing CCD observations, by member James Screech, of the eclipsing variable AC Boötis during the night of 15/16 April 2020. The observations were made using an Atik 414ex mono camera, photometric V filter and an ED70 refractor mounted on a Celestron AVX mount. (BAA Variable Star Section)

Light Curve for Z CAM

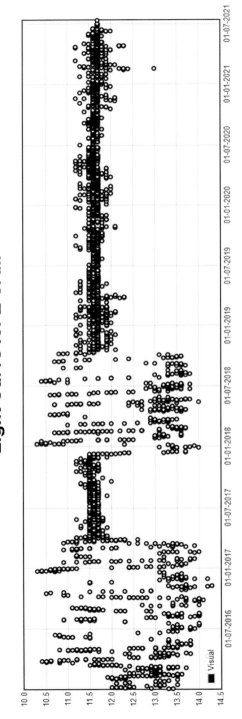

The BAA VSS light curve for Z Camelopardalis between 2016 and mid-2021. The ups and downs are due to short-lived dwarf nova outbursts, but note also the significant 'standstills' in brightness near magnitude 11.5 (BAA Variable Star Section)

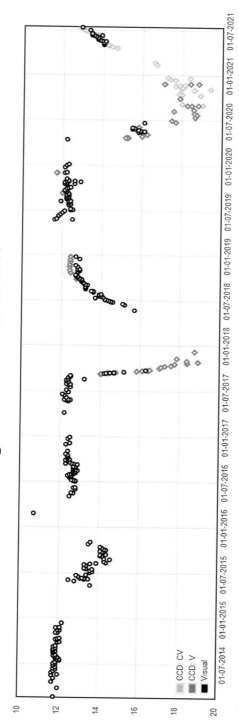

Light Curve for V742 LYR

The BAA VSS light curve for the R Coronae Borealis (RCB) type variable V742 Lyrae from 2014 to mid-2021. (BAA Variable Star Section)

Light Curve for DY PER

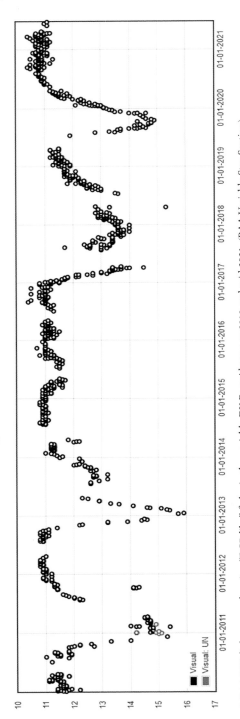

A BAA VSS light curve showing "RCB-like" fades in the variable DY Persei between 2010 and mid-2021. (BAA Variable Star Section)

Light Curve for W AND

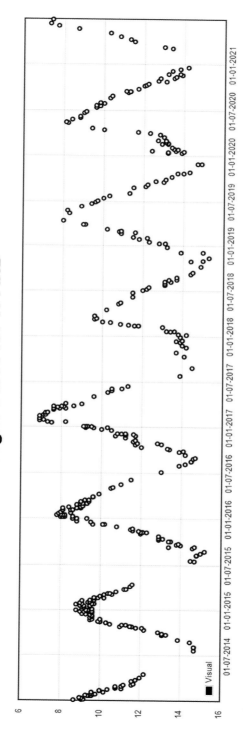

The BAA VSS light curve for the Mira type variable W Andromedae from 2014 to mid-2021. Note how the brightness at maximum has varied significantly from cycle to cycle. (BAA Variable Star Section)

Light Curve for BQ ORI

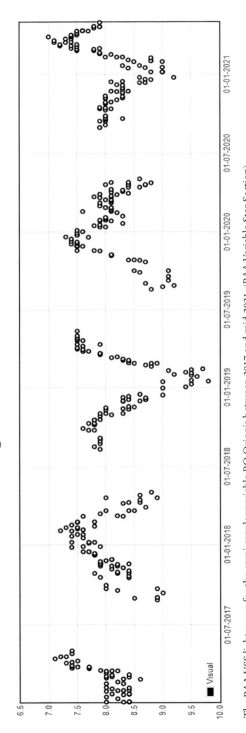

The BAA VSS light curve for the semi-regular variable BQ Orionis between 2017 and mid-2021. (BAA Variable Star Section)

Some Interesting Double Stars

Brian Jones

The accompanying table describes the visual appearances of a selection of double stars. These may be optical doubles (which consist of two stars which happen to lie more or less in the same line of sight as seen from Earth and which therefore only appear to lie close to each other) or binary systems (which are made up of two stars which are gravitationally linked and which orbit their common centre of mass).

Other than the location on the celestial sphere and the magnitudes of the individual components, the list gives two other values for each of the double stars listed – the angular separation and position angle (PA). Further details of what these terms mean can be found in the article *Double and Multiple Stars* published in the 2018 edition of the Yearbook of Astronomy.

Double-star observing can be a very rewarding process, and even a small telescope will show most, if not all, the best doubles in the sky. You can enjoy looking at double stars simply for their beauty, such as Albireo (β Cygni) or Almach (γ Andromedae), although there is a challenge to be had in splitting very difficult (close) double stars, such as the demanding Sirius (α Canis Majoris) or the individual pairs forming the Epsilon (ε) Lyrae 'Double-Double' star system.

The accompanying list is a compilation of some of the prettiest double (and multiple) stars scattered across both the Northern and Southern heavens. Once you have managed to track these down, many others are out there awaiting your attention …

Star	RA		Declination		Magnitudes	Separation	PA	Comments
	h	m	°	'		(arcsec)	°	
Beta[1,2] (β[1,2]) Tucanae	00	31.5	−62	58	4.36 / 4.53	27.1	169	Both stars again double, but difficult
Achird (η Cassiopeiae)	00	49.1	+57	49	3.44 / 7.51	13.4	324	Easy double
Mesarthim (γ Arietis)	01	53.5	+19	18	4.58 / 4.64	7.6	1	Easy pair of white stars
Almach (γ Andromedae)	02	03.9	+42	20	2.26 / 4.84	9.6	63	Yellow and blue-green components
32 Eridani	03	54.3	−02	57	4.8 / 6.1	6.9	348	Yellowish and bluish
Alnitak (ζ Orionis)	05	40.7	−01	57	2.0 / 4.3	2.3	167	Difficult, can be resolved in 10cm telescopes
Gamma (γ) Leporis	05	44.5	−22	27	3.59 / 6.28	95.0	350	White and yellow-orange components, easy pair
Sirius (α Canis Majoris)	06	45.1	−16	43	−1.4 / 8.5			Binary, period 50 years, difficult
Castor (α Geminorum)	07	34.5	+31	53	1.93 / 2.97	7.0	55	Binary, 445 years, widening
Gamma (γ) Velorum	08	09.5	−47	20	1.83 / 4.27	41.2	220	Pretty pair in nice field of stars
Upsilon (υ) Carinae	09	47.1	−65	04	3.08 / 6.10	5.03	129	Nice object in small telescopes
Algieba (γ Leonis)	10	20.0	+19	50	2.28 / 3.51	4.6	126	Binary, 510 years, orange-red and yellow
Acrux (α Crucis)	12	26.4	−63	06	1.40 / 1.90	4.0	114	Glorious pair, third star visible in low power
Porrima (γ Virginis)	12	41.5	−01	27	3.56 / 3.65			Binary, 170 years, widening, visible in small telescopes
Cor Caroli (α Canum Venaticorum)	12	56.0	+38	19	2.90 / 5.60	19.6	229	Easy, yellow and bluish
Mizar (ζ Ursae Majoris)	13	24.0	+54	56	2.3 / 4.0	14.4	152	Easy, wide naked-eye pair with Alcor
Alpha (α) Centauri	14	39.6	−60	50	0.0 / 1.2			Binary, beautiful pair of stars
Izar (ε Boötis)	14	45.0	+27	04	2.4 / 5.1	2.9	344	Fine pair of yellow and blue stars
Omega[1,2] (ω[1,2]) Scorpii	16	06.0	−20	41	4.0 / 4.3	14.6	145	Optical pair, easy
Epsilon[1] (ε[1]) Lyrae	18	44.3	+39	40	4.7 / 6.2	2.6	346	The Double-Double, quadruple system with ε[2]
Epsilon[2] (ε[2]) Lyrae	18	44.3	+39	40	5.1 / 5.5	2.3	76	Both individual pairs just visible in 80mm telescopes
Theta[1,2] (θ[1,2]) Serpentis	18	56.2	+04	12	4.6 / 5.0	22.4	104	Easy pair, mag 6.7 yellow star 7 arc minutes from θ[2]

Star	RA		Declination		Magnitudes	Separation	PA	Comments
	h	m	°	'		(arcsec)	°	
Albireo (β Cygni)	19	30.7	+27	58	3.1 / 5.1	34.3	54	Glorious pair, yellow and blue-green
Algedi (α[1,2] Capricorni)	20	18.0	−12	32	3.7 / 4.3	6.3	292	Optical pair, easy
Gamma (γ) Delphini	20	46.7	+16	07	5.14 / 4.27	9.2	265	Easy, orange and yellow-white
61 Cygni	21	06.9	+38	45	5.20 / 6.05	31.6	152	Binary, 678 years, both orange
Theta (θ) Indi	21	19.9	−53	27	4.6 / 7.2	7.0	275	Fine object for small telescopes
Delta (δ) Tucanae	22	27.3	−64	58	4.49 / 8.7	7.0	281	Beautiful double, white and reddish

Some Interesting Nebulae, Star Clusters and Galaxies

Brian Jones

Object	RA		Declination		Remarks
	h	m	°	'	
47 Tucanae (in Tucana)	00	24.1	−72	05	Fine globular cluster, easy with naked eye
M31 (in Andromeda)	00	40.7	+41	05	Andromeda Galaxy, visible to unaided eye
Small Magellanic Cloud	00	52.6	−72	49	Satellite galaxy of the Milky Way
NGC 362 (in Tucana)	01	03.3	−70	51	Globular cluster, impressive sight in telescopes
M33 (in Triangulum)	01	31.8	+30	28	Triangulum Spiral Galaxy, quite faint
NGC 869 and NGC 884	02	20.0	+57	08	Sword Handle Double Cluster in Perseus
M34 (in Perseus)	02	42.1	+42	46	Open star cluster near Algol
M45 (in Taurus)	03	47.4	+24	07	Pleiades or Seven Sisters cluster, a fine object
Large Magellanic Cloud	05	23.5	−69	45	Satellite galaxy of the Milky Way
30 Doradus (in Dorado)	05	38.6	−69	06	Star-forming region in Large Magellanic Cloud
M1 (in Taurus)	05	32.3	+22	00	Crab Nebula, near Zeta (ζ) Tauri
M38 (in Auriga)	05	28.6	+35	51	Open star cluster
M42 (in Orion)	05	33.4	−05	24	Orion Nebula
M36 (in Auriga)	05	36.2	+34	08	Open star cluster
M37 (in Auriga)	05	52.3	+32	33	Open star cluster
M35 (in Gemini)	06	06.5	+24	21	Open star cluster near Eta (η) Geminorum
M41 (in Canis Major)	06	46.0	−20	46	Open star cluster to south of Sirius
M44 (in Cancer)	08	38.0	+20	07	Praesepe, visible to naked eye
M81 (in Ursa Major)	09	55.5	+69	04	Bode's Galaxy
M82 (in Ursa Major)	09	55.9	+69	41	Cigar Galaxy or Starburst Galaxy
Carina Nebula (in Carina)	10	45.2	−59	52	NGC 3372, large area of bright and dark nebulosity
M104 (in Virgo)	12	40.0	−11	37	Sombrero Hat Galaxy to south of Porrima
Coal Sack (in Crux)	12	50.0	−62	30	Prominent dark nebula, visible to naked eye
NGC 4755 (in Crux)	12	53.6	−60	22	Jewel Box open cluster, magnificent object
Omega (ω) Centauri	13	23.7	−47	03	Splendid globular in Centaurus, easy with naked eye
M51 (in Canes Venatici)	13	29.9	+47	12	Whirlpool Galaxy
M3 (in Canes Venatici)	13	40.6	+28	34	Bright Globular Cluster

Object	RA h	RA m	Declination °	Declination '	Remarks
M4 (in Scorpius)	16	21.5	−26	26	Globular cluster, close to Antares
M12 (in Ophiuchus)	16	47.2	−01	57	Globular cluster
M10 (in Ophiuchus)	16	57.1	−04	06	Globular cluster
M13 (in Hercules)	16	40.0	+36	31	Great Globular Cluster, just visible to naked eye
M92 (in Hercules)	17	16.1	+43	11	Globular cluster
M6 (in Scorpius)	17	36.8	−32	11	Open cluster
M7 (in Scorpius)	17	50.6	−34	48	Bright open cluster
M20 (in Sagittarius)	18	02.3	−23	02	Trifid Nebula
M8 (in Sagittarius)	18	03.6	−24	23	Lagoon Nebula, just visible to naked eye
M16 (in Serpens)	18	18.8	−13	49	Eagle Nebula and star cluster
M17 (in Sagittarius)	18	20.2	−16	11	Omega Nebula
M11 (in Scutum)	18	49.0	−06	19	Wild Duck open star cluster
M57 (in Lyra)	18	52.6	+32	59	Ring Nebula, brightest planetary
M27 (in Vulpecula)	19	58.1	+22	37	Dumbbell Nebula
M29 (in Cygnus)	20	23.9	+38	31	Open cluster
M15 (in Pegasus)	21	28.3	+12	10	Bright globular cluster near Epsilon (ε) Pegasi
M39 (in Cygnus)	21	31.6	+48	25	Open cluster, good with low powers
M52 (in Cassiopeia)	23	24.2	+61	35	Open star cluster near 4 Cassiopeiae

M = Messier Catalogue Number   NGC = New General Catalogue Number

The positions in the sky of each of the objects contained in this list are given on the Monthly Star Charts printed elsewhere in this volume.

Astronomical Organizations

American Association of Variable Star Observers

49 Bay State Road, Cambridge, Massachusetts, 02138, USA

aavso.org

The AAVSO is an international non-profit organization of variable star observers whose mission is to enable anyone, anywhere, to participate in scientific discovery through variable star astronomy. We accomplish our mission by carrying out the following activities:

- observation and analysis of variable stars
- collecting and archiving observations for worldwide access
- forging strong collaborations between amateur and professional astronomers
- promoting scientific research, education and public outreach using variable star data

American Astronomical Society

1667 K Street NW, Suite 800, Washington, DC 20006, USA

aas.org

Established in 1899, the American Astronomical Society (AAS) is the major organization of professional astronomers in North America. The mission of the AAS is to enhance and share humanity's scientific understanding of the universe, which it achieves through publishing, meeting organization, education and outreach, and training and professional development.

Association of Lunar and Planetary Observers (ALPO)

Matthew L. Will (Secretary), P.O. Box 13456, Springfield, IL 62791-3456, USA

alpo-astronomy.org

Founded in 1947 by Walter Haas, the ALPO is an international non-profit organization that studies all natural bodies in our solar system. ALPO Sections include Lunar, Solar, Mercury, Venus, Mars, Minor Planets, Jupiter, Saturn, Remote Planets, Comets, Meteors, Meteorites, Eclipses, Exoplanets, Outreach and Online, many with separate "Studies Programs" within these Sections. Minimum membership is very reasonable and includes the quarterly full colour digital *Journal of the ALPO*. Interested observers of any experience are welcome to join. Many members stand ready to improve the skills and abilities of novices.

Astronomical Society of the Pacific

390 Ashton Avenue, San Francisco, CA 94112, USA
astrosociety.org

Formed in 1889, the Astronomical Society of the Pacific (ASP) is a non-profit membership organization which is international in scope. The mission of the ASP is to increase the understanding and appreciation of astronomy through the engagement of our many constituencies to advance science and science literacy. We invite you to explore our site to learn more about us; to check out our resources and education section for the researcher, the educator, and the backyard enthusiast; to get involved by becoming an ASP member; and to consider supporting our work for the benefit of a science literate world!

Astrospeakers.org

astrospeakers.org

A website designed to help astronomical societies and clubs locate astronomy and space lecturers which is also designed to help people find their local astronomical society. It is completely free to register and use and, with over 50 speakers listed, is an excellent place to find lecturers for your astronomical society meetings and events. Speakers and astronomical societies are encouraged to use the online registration to be added to the lists.

British Astronomical Association

Burlington House, Piccadilly, London, W1J 0DU, England
britastro.org

The British Astronomical Association is the UK's leading society for amateur astronomers catering for beginners to the most advanced observers who produce scientifically useful observations. Our Observing Sections provide encouragement and advice about observing. We hold meetings around the country and publish a bi-monthly Journal plus an annual Handbook. For more details, including how to join the BAA or to contact us, please visit our website.

British Interplanetary Society

Arthur C. Clarke House, 27/29 South Lambeth Road, London, SW8 1SZ, England
bis-space.com
The British Interplanetary Society is the world's longest-established space advocacy organisation, founded in 1933 by the pioneers of British astronautics. It is the first organisation in the world still in existence to design spaceships. Early members included Sir Arthur C Clarke and Sir Patrick Moore. The Society has created many original concepts, from a 1938 lunar lander and space suit designs, to geostationary orbits, space stations and the first engineering study of a starship, Project Daedalus. Today the BIS has a worldwide membership and welcomes all with an interest in Space, including enthusiasts, students, academics and professionals.

Canadian Astronomical Society
Société Canadienne D'astronomie (CASCA)

100 Viaduct Avenue West, Victoria, British Columbia, V9E 1J3, Canada
casca.ca
CASCA is the national organization of professional astronomers in Canada. It seeks to promote and advance knowledge of the universe through research and education. Founded in 1979, members include university professors, observatory scientists, postdoctoral fellows, graduate students, instrumentalists, and public outreach specialists.

Royal Astronomical Society of Canada

203-4920 Dundas St W, Etobicoke, Toronto, ON M9A 1B7, Canada
rasc.ca
Bringing together over 5,000 enthusiastic amateurs, educators and professionals RASC is a national, non-profit, charitable organization devoted to the advancement of astronomy and related sciences and is Canada's leading astronomy organization. Membership is open to everyone with an interest in astronomy. You may join through any one of our 29 RASC centres, located across Canada and all of which offer local programs. The majority of our events are free and open to the public.

Federation of Astronomical Societies

The Secretary, 147 Queen Street, Swinton, Mexborough, S64 8NG
fedastro.org.uk
The Federation of Astronomical Societies (FAS) is an umbrella group for astronomical societies in the UK. It promotes cooperation, knowledge and

information sharing and encourages best practice. The FAS aims to be a body of societies united in their attempts to help each other find the best ways of working for their common cause of creating a fully successful astronomical society. In this way it endeavours to be a true federation, rather than some remote central organization disseminating information only from its own limited experience. The FAS also provides a competitive Public Liability Insurance scheme for its members.

International Dark-Sky Association
darksky.org

The International Dark-Sky Association (IDA) is the recognized authority on light pollution and the leading organization combating light pollution worldwide. The IDA works to protect the night skies for present and future generations, our public outreach efforts providing solutions, quality education and programs that inform audiences across the United States of America and throughout the world. At the local level, our mission is furthered through the work of our U.S. and international chapters representing five continents.

The goals of the IDA are:

- Advocate for the protection of the night sky
- Educate the public and policymakers about night sky conservation
- Promote environmentally responsible outdoor lighting
- Empower the public with the tools and resources to help bring back the night

The Planetary Society
60 South Los Robles Avenue, Pasadena, CA 91101, USA
planetary.org

The Planetary Society was founded by Carl Sagan, Louis Friedman and Bruce Murray in 1980 in direct response to the enormous public interest in space, and with a mission to introduce people to the wonders of the cosmos. With a global membership in excess of 50,000 from over 100 countries, it is the largest and most influential non-profit space organization in the world. The Planetary Society bridges the gap between the scientific community and the general public, inspiring and educating people from all walks of life and empowering the world's citizens to advance space science and exploration.

Royal Astronomical Society of New Zealand

PO Box 3181, Wellington, New Zealand

rasnz.org.nz

Founded in 1920, the object of The Royal Astronomical Society of New Zealand is the promotion and extension of knowledge of astronomy and related branches of science. It encourages interest in astronomy and is an association of observers and others for mutual help and advancement of science. Membership is open to all interested in astronomy. The RASNZ has about 180 individual members including both professional and amateur astronomers and many of the astronomical research and observing programmes carried out in New Zealand involve collaboration between the two. In addition the society has a number of groups or sections which cater for people who have interests in particular areas of astronomy.

Astronomical Society of Southern Africa

Astronomical Society of Southern Africa, c/o SAAO, PO Box 9, Observatory, 7935, South Africa

assa.saao.ac.za

Formed in 1922, The Astronomical Society of Southern Africa comprises both amateur and professional astronomers. Membership is open to all interested persons. Regional Centres host regular meetings and conduct public outreach events, whilst national Sections coordinate special interest groups and observing programmes. The Society administers two Scholarships, and hosts occasional Symposia where papers are presented. For more details, or to contact us, please visit our website.

Royal Astronomical Society

Burlington House, Piccadilly, London, W1J 0BQ, England

ras.org.uk

The Royal Astronomical Society, with around 4,000 members, is the leading UK body representing astronomy, space science and geophysics, with a membership including professional researchers, advanced amateur astronomers, historians of science, teachers, science writers, public engagement specialists and others.

Society for Popular Astronomy

Secretary: Guy Fennimore, 36 Fairway, Keyworth, Nottingham, NG12 5DU

popastro.com

The Society for Popular Astronomy is a national society that aims to present astronomy in a less technical manner. The bi-monthly society magazine Popular Astronomy is issued free to all members.

Our Contributors

Martin Braddock is a professional scientist and project leader in the field of drug discovery and development with 36 years' experience of working in academic institutes and large corporate organizations. He holds a BSc in Biochemistry and a PhD in Radiation Biology and is a former Royal Society University Research Fellow at the University of Oxford. He was elected a Fellow of the Royal Society of Biology in 2010, and in 2012 received an Alumnus Achievement Award for distinction in science from the University of Salford. Martin has published over 190 peer-reviewed scientific publications, filed nine patents, and edited two books for the Royal Society of Chemistry and serves as a grant proposal evaluator for multiple international research agencies. Martin holds further qualifications from the University of Central Lancashire and Open University. He is a member of the Mansfield and Sutton Astronomical Society and was elected a Fellow of the Royal Astronomical Society in May 2015 and has given over 100 lectures to astronomical societies in the U.K. An ambassador for science, technology, engineering and mathematics (STEM), Martin seeks to inspire the next generation of young scientists to aim high and be the best they can be. To find out more about him visit **science4u.co.uk**

Neil Haggath has a degree in astrophysics from Leeds University and has been a Fellow of the Royal Astronomical Society since 1993. A member of Cleveland and Darlington Astronomical Society since 1981, he has served on its committee since 1989. Neil is an avid umbraphile, clocking up six total eclipse expeditions so far to locations as far flung as Australia and Hawai'i. Four of them were successful, the most recent being in Jackson, Wyoming on 21 August 2017. In 2012, he may have set a somewhat unenviable record among British astronomers - for the greatest distance travelled (6,000 miles to Thailand) to NOT see the transit of Venus. He saw nothing on the day . . . and got very wet!

Dr. David M. Harland gained his BSc in astronomy in 1977, lectured in computer science, worked in industry, and managed academic research. In 1995 he 'retired' in order to write on space themes.

David Harper, FRAS has had a varied career which includes teaching mathematics, astronomy and computing at Queen Mary University of London, astronomical software development at the Royal Greenwich Observatory, bioinformatics support at the Wellcome Trust Sanger Institute, and a research interest in the dynamics of planetary satellites, which began during his Ph.D. at Liverpool University in the 1980s and continues in an occasional collaboration with colleagues in China. He is married to fellow contributor Lynne Stockman.

Tracie Heywood is an amateur astronomer from Leek in Staffordshire and is one of the UK's leading variable star observers, using binoculars to monitor the brightness changes of several hundred variable stars. Tracie currently writes a monthly column about variable stars for *Astronomy Now* magazine. She has previously been the Eclipsing Binary coordinator for the Variable Star Section of the British Astronomical Association and the Director of the Variable Star Section of the Society for Popular Astronomy.

Rod Hine was aged around ten when he was given a copy of *The Boys Book of Space* by Patrick Moore. Already interested in anything to do with science and engineering he devoured the book from cover to cover. The launch of Sputnik I shortly afterwards clinched his interest in physics and space travel. He took physics, chemistry and mathematics at A-level and then studied Natural Sciences at Churchill College, Cambridge. He later switched to Electrical Sciences and subsequently joined Marconi at Chelmsford working on satellite communications in the UK, Middle East and Africa. This led to work in meteorological communications in Nairobi, Kenya and later a teaching post at the Kenya Polytechnic. There he met and married a Yorkshire lass and moved back to the UK in 1976. Since then he has had a variety of jobs in electronics and industrial controls, and until recently was lecturing part-time at the University of Bradford. Rod got fully back into astronomy in around 1992 when his wife bought him an astronomy book, at which time he joined Bradford Astronomical Society. With the call sign G8AQH, he is also a member of Otley Amateur Radio Society. He is currently working part-time at Leeds University providing engineering support for a project to convert redundant satellite dishes into radio telescopes in developing countries.

Brian Jones hails from Bradford in the West Riding of Yorkshire and was a founder member of the Bradford Astronomical Society. He developed a fascination with astronomy at the age of five when he first saw the stars through a pair of binoculars, although he spent the first part of his working life developing a career in mechanical engineering. However, his true passion lay in the stars and his interest in astronomy took him into the realms of writing sky guides for local newspapers, appearing on local radio and television, teaching astronomy and space in schools and, in 1985, leaving engineering to become a full time astronomy and space writer. His books have covered a range of astronomy and space-related topics for both children and adults and his journalistic work includes writing articles and book reviews for several astronomy magazines as well as for many general interest magazines, newspapers and periodicals. His passion for bringing an appreciation of the universe to his readers is reflected in his writing. You can follow Brian on Twitter via **@StarsBrian** and check out the sky by visiting his blog at **starlight-nights.co.uk** from where you can also access his Facebook group Starlight Nights.

Carolyn Kennett lives in the far south-west of Cornwall. She likes to write, although you will often find her walking the Tinner's Way and coastal pathways found in the local countryside. As well as researching local astronomy history, she has a passion for archaeoastronomy and eighteenth- and nineteenth-century astronomy.

Andrew P.B. Lound is a lecturer, writer and broadcaster with a background in science, history, engineering and performing arts and has spent over forty years promoting science to the general public. He worked with The Planetary Society for thirty-five years managing public awareness events and science expeditions as well as promoting international cooperation – winning a number of awards in the process. He is the former curator of the Avery Historical Museum and has worked at a number of historic sites. Andrew has published four books covering the history of science and industry and can be regularly heard on Talk Radio commenting on space science matters; BBC Radio where he covers space science and history; and can be seen on GBNews commenting on space science topics. He has also written and presented a number of television programmes and has performed in several historic acting roles. He regularly tours with his trademark Odyssey Dramatic Presentations. To find out more about Andrew visit **andrewlound.com**

Harold A. McAlister is Regents Professor Emeritus of Astronomy at Georgia State University where he was founder and first Director of the Center for High Angular Resolution Astronomy which built and operates the CHARA Array on Mount Wilson, California. From 2002 to 2014 he was CEO of the Mount Wilson Institute and Director of Mount Wilson Observatory. In his retirement McAlister writes fiction and non-fiction books, the most recent of which is *Seeing the Unseen: Mount Wilson's Role in High Angular Resolution Astronomy*. For more, see **halmcalister.com**

John McCue graduated in astronomy from the University of St Andrews and began teaching. He gained a Ph.D. from Teesside University studying the unusual rotation of Venus. In 1979 he and his colleague John Nichol founded the Cleveland and Darlington Astronomical Society, which then worked in partnership with the local authority to build the Wynyard Planetarium and Observatory in Stockton-on-Tees. John is currently double star advisor for the British Astronomical Association.

Mary McIntyre is an amateur astronomer, astrophotographer and astronomy sketcher based in Oxfordshire, UK. She has had a life-long interest in astronomy and loves sharing that with others. Mary is a writer and communicator and has contributed several articles to *Sky at Night* magazine and made numerous radio appearances including being a co-presenter of the monthly radio show Comet Watch and a member of the AstroRadio team. She also gives regular astronomy and photography talks to people of all ages, including presentations to astronomy societies, camera clubs, schools and Beavers/Cubs/Scouts groups. She is passionate about promoting science, technology, engineering and mathematics (STEM) and astronomy to women and girls and runs the UK Women in Astronomy Network, which connects, celebrates and promotes women with a passion for astronomy. By providing role models from both professional and the keen amateur, women and young girls are encouraged and supported to explore all aspects of astronomy and astrophotography and given the confidence to pursue a career in the field. To find out more about Mary visit **marymcintyreastronomy.co.uk**

Neil Norman, FRAS first became fascinated with the night sky when he was five years of age and saw Patrick Moore on the television for the first time. It was the Sky at Night programme, broadcast in March 1986 and dedicated to the Giotto probe reaching Halley's Comet, which was to ignite his passion for these icy interlopers. As the years passed, he began writing astronomy articles for local news magazines before moving into internet radio where he initially guested on the Astronomyfm show 'Under British Skies', before becoming a co-host for a short time. In 2013 he created Comet Watch, a Facebook group dedicated to comets of the past, present and future. His involvement with Astronomyfm led to the creation of the monthly radio show Comet Watch, which is now in its fourth year. Neil lives in Suffolk with his partner and three children. Perhaps rather fittingly, given Neil's interest in asteroids, he has one named in his honour, this being the main belt asteroid 314650 Neilnorman, discovered in July 2006 by English amateur astronomer Matt Dawson.

Peter Rea has had a keen interest in lunar and planetary exploration since the early 1960s and frequently lectures on the subject. He helped found the Cleethorpes and District Astronomical Society in 1969. In April of 1972 he was at the Kennedy Space Centre in Florida to see the launch of Apollo 16 to the moon and in October 1997 was at the southern end of Cape Canaveral to see the launch of Cassini to Saturn. He would still like to see a total solar eclipse as the expedition he was on to see the 1973 eclipse in Mali had vehicle trouble and the meteorologists decided he was not going to see the 1999 eclipse from Devon. He lives in Lincolnshire with his wife Anne and has a daughter who resides in New Zealand.

Richard H. Sanderson retired in 2018 after a 19-year tenure as curator of physical science at the Springfield Science Museum in Massachusetts, where he managed the Seymour Planetarium, the oldest operating projection planetarium in the United States. Richard wrote a newspaper astronomy column for many years and co-authored the 2006 book *Illustrated Timeline of the Universe*. He has served as president of the Springfield Stars Club and helps organize the Connecticut River Valley Astronomers' Conjunction, an annual convention of astronomy enthusiasts. His passion for antiquarian astronomy books and memorabilia has flourished for more than 40 years and led to his creation of the "Vintage Astronomy Books" Facebook group, located at **facebook.com/groups/vintageastronomy**

Lynne Marie Stockman holds degrees in mathematics from Whitman College, the University of Washington and the University of London. She has studied astronomy at both undergraduate and postgraduate levels, and is a member of the Astronomical Society of the Pacific. A native of North Idaho, Lynne has lived in Britain since 1992. She was an early pioneer of the World Wide Web: with her husband and fellow Yearbook contributor David Harper, she created the web site **obliquity.com** in 1998 to share their interest in astronomy, computing, family history and cats.

Become a Friend
of the
Royal Astronomical
Society

Benefits include...

Friends of the RAS lecture programme
Visits to observatories and other places of interest
Discounted subscription to 'Astronomy & Geophysics' magazine
Use of the RAS library

For more information visit:

www.ras.ac.uk/friends

Royal
Astronomical
Society

DEFENDING EARTH FROM ASTEROIDS

EXPLORING THE COSMOS

SEARCHING FOR LIFE
ON OTHER WORLDS

GOOD IDEAS, RIGHT?

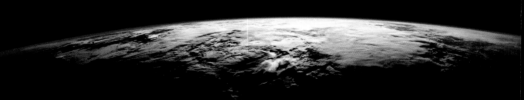

WE THINK SO TOO.

Join the most effective space nonprofit on this planet at planetary.org

Appetite whetted?

Join the UK's leading organisation
for observers of the night sky

British Astronomical Association

Burlington House, Piccadilly, London W1J 0DU
Tel: 020 7734 4145
Email: office@britastro.org

visit www.britastro.org

Royal Astronomical Society of New Zealand

---

22 affiliated community societies

Experts in southern hemisphere skies

Strong collaborations between professional & amateur observers

---

We welcome all interested in astronomy

www.rasnz.org.nz

Support all year round!

The Society for Popular Astronomy is for YOU

The Society for Popular Astronomy is a UK-based organisation that is here to help beginners. It is a role we've been playing since 1953 when we were founded by Sir Patrick Moore as the Junior Astronomical Society. Yes, we're even older than this *Yearbook*!

The SPA has moved with the times over the many years since we began, but our essential purpose remains the same – to offer affordable membership that gives beginners of all ages advice and support in their interest in astronomy and space. We believe in "Stargazing for everyone". Why not come and join us! Visit **www.popastro.com** to see what we offer.

Society for Popular Astronomy
STARGAZING FOR EVERYONE

- **Popular Astronomy – our great magazine**
- **Helpful website plus a forum for chat**
- **Advice and useful observing guides**
- **Meetings, courses, workshops and videos**
- **Special rates for Young Stargazers**

www.popastro.com

See our range of T-shirts, fun mugs and astro merchandise!

The Commission for Dark Skies (CfDS) was founded in 1989 by concerned members of the British Astronomical Association. It works towards better night skies for all and the protection of the terrestrial environment and biodiversity from the depredations of wasted artificial light.

It's not difficult to define light pollution. It's light that's wasted, goes to the wrong place and isn't lighting only the places that need to be lit. Nobody wants to switch off all lights and live in medieval darkness. The solution is "the right light in the right place at the right time".

Light pollution has increased year on year for several decades, and modern bright LED lights are not helping. Most are too bright for the job and too white (blue-rich), damaging the environment and often intruding into nearby premises to cause light nuisance.

The night sky, half of our visual environment, has no protection in law in the UK; despoiling the environment down here, by fly-tipping for example, can carry stiff penalties and even land offenders in prison. But spoiling the beauty of the night sky is permitted. Light pollution is not just an astronomer's problem. A lot of energy and money is wasted by ill-directed lights. Stray light, fraying the edges of the habitat of night, contributes to the current biodiversity collapse (most invertebrate species are nocturnal), accelerates climate change and harms our health as our ancient circadian rhythms are disrupted.

Most people don't realise the harm light pollution causes. Everyone has grown up with lighting, public and private, that is glary, over-bright and shines where it shouldn't. It seems normal.

To support the work of the Commission for Dark Skies, contact us at https://britastro.org/dark-skies/enquiries.php

The CfDS works closely with other like-minded groups in many countries, for example **the International Dark-Sky Association** (www.darksky.org).

It also maintains close links with the **All-Party Parliamentary Dark Skies Group of MPs** https://appgdarkskies.co.uk

The Federation of Astronomical Societies

The FAS is a union of astronomical societies, groups and clubs, liaising together, where practicable, for their mutual benefit.

Formed in 1974, we shall be celebrating our **50th anniversary** in 2024.

Watch out for announcements of events to mark this milestone, including webinars and live conventions.

For details about the organisation, as well as other information, visit our website at **www.fedastro.org.uk**

1974 - 2024

Benefits of membership include:

- Access to Public Liability Insurance for societies at extremely favourable rates
- A Resource Centre containing documents providing a common framework for the running of astronomical societies and their outreach activities
- A regular Newsletter where societies can feature their events and information
- Regular Conventions featuring prominent speakers from Academic and Industrial backgrounds at sites around the UK, and online webinars and lectures
- A website announcing events from societies and other organisations that would be of interest to its member societies
- More than 200 astronomical societies are FAS Members, not only in the United Kingdom but also overseas.

Matt Nicholls addressing our convention at the National Space Centre, Leicester

Photo by William Bottaci